EVALUATION OF FEASIBILITY OF METHODS TO MINIMIZE BIOMASS PRODUCTION FROM BIOTREATMENT

by:
H. David Stensel and Stuart E. Strand
University of Washington

2004

WERF

The Water Environment Research Foundation, a not-for-profit organization, funds and manages water quality research for its subscribers through a diverse public-private partnership between municipal utilities, corporations, academia, industry, and the federal government. WERF subscribers include municipal and regional water and wastewater utilities, industrial corporations, environmental engineering firms, and others that share a commitment to cost-effective water quality solutions. WERF is dedicated to advancing science and technology addressing water quality issues as they impact water resources, the atmosphere, the lands, and quality of life.

For more information, contact:
Water Environment Research Foundation
635 Slaters Lane, Suite 300
Alexandria, VA 22314-1177
Tel: (703) 684-2470
Fax: (703) 299-0742
www.werf.org
werf@werf.org

This report was co-published by the following organizations. For non-subscriber sales information, contact:

IWA Publishing
Alliance House, 12 Caxton Street
London SW1H 0QS, United Kingdom
Tel: +44 (0) 20 7654 5500
Fax: +44 (0) 20 7654 5555
www.iwapublishing.com
publications@iwap.co.uk

Library of Congress Catalog Card Number: 2003111577

IWAP ISBN: 1-84339-703-X

University of Washington

The research on which this report is based was funded, in part, by the United States Environmental Protection Agency through Cooperative Agreement No. CR-827345-01 with the Water Environment Research Foundation (WERF). Unless an EPA logo appears on the cover, this report is a publication of WERF, not EPA. Funds awarded under the Cooperative Agreement cited above were not used for editorial services, reproduction, printing, or distribution.

ACKNOWLEDGMENTS

The project team would also like to acknowledge and thank the laboratory and operations staff at the King County West Point wastewater treatment facility, which made the field data collection part of this work possible. We also thank Dr. Mohammad Abu-Orf and Veolia Water North America for making the membrane separation pilot plant available for testing at the West Point wastewater treatment facility. We also thank Wenjie Lu and Xing Zhang, visiting students from Sichuan University, China for their laboratory work at the University of Washington.

Report Preparation

Principal Investigators:
H. David Stensel, Ph.D., P.E.
University of Washington

Stuart E. Strand, Ph.D.
University of Washington

Project Team:
Greg Bowden
Veolia Water North America

Robert Bucher
King County Department of Natural Resources

Glen T. Daigger, Ph.D., P.E.
CH2M HILL

Rebecca Moore
Moore Engineering Consultants

Project Subcommittee

John B. Barber, Ph.D., Chair
Eastman Chemical

Robert A. Reich, P.E.
Dupont Company

Richard C. Butler
King County WPCD East

Robert A. Gillette, P.E., DEE
Carollo Engineers

ABSTRACT AND BENEFITS

Abstract:

This report presents the results of an evaluation of technologies that may result in less biomass production in activated sludge processes. The report summarizes the results of a comprehensive literature review that was done to evaluate technologies in terms of their sludge reduction potential, ease of implementation, impacts on plant operations and effluent quality, reliability, and relative capital and operating costs.

Reporting testing results supported significant biomass reduction by processes using chemical and thermal methods, higher life forms (predator processes), anaerobic instead of aerobic respiration, and extreme solids retention times, but biomass reduction for enhanced biological phosphorus removal (EBPR) processes and a mechanical disintegration process were less conclusive. The predator enhancement process showed promise for industrial wastewater treatment, but is less attractive for municipal wastewater treatment for which a lower soluble COD fraction is present. Extreme solids retention time processes may be practical for small wastewater flows and perhaps with the use of membrane separation technology. Anaerobic treatment processes are known to have a lower biomass yield (one fourth or a less than for aerobic treatment), but work is needed to develop their applications for low strength, low temperature wastewaters, such as in municipal wastewater treatment. For some processes such as the cell disruption using mechanical, thermal, and chemical means, the cost of implementing the biomass reduction technology was greater than the cost savings associated with less sludge production. Addition of chemical uncouplers can greatly reduce biomass production, but pose problems of toxic chemicals in the treated effluent.

In a series of bench-scale tests carried out at the Seattle West Point wastewater treatment facility and the University of Washington environmental engineering laboratories the presence and mechanism of COD loss (and subsequent less biomass production) in the anaerobic zone of EBPR processes was investigated. The results of the test work and fundamental evaluation could not support previous claims of a COD loss in EBPR processes, nor was less sludge production observed.

Benefits:

- Describes and evaluates a wide range of technologies that can reduce biomass sludge production.
- Shows experience with anaerobic treatment processes treating domestic or industrial wastewaters at temperatures from 10–30 ^{0}C.
- Indicates most promising technologies for biomass reduction in activated sludge systems.
- Shows factors that affect total sludge production in activated sludge treatment and how application of sludge reduction technologies can result in increased wastewater treatment plant capacity.
- Provides additional information that may affect process selection in wastewater treatment process alternative assessments.

- Shows that assumptions for COD loss in anaerobic/aerobic biological phosphorus removal systems may lead to plants that are under-designed in terms of oxygen requirements and sludge production.
- Identifies needed research to further exploit potential biomass reduction technologies.

Keywords: Activated sludge, anaerobic, aerobic, COD loss, membrane separation, mechanical disintegration, ozonation, predators, sludge production, uncouplers

TABLE OF CONTENTS

LIST OF TABLES

LIST OF FIGURES

LIST OF ACRONYMS

The following provides definitions for the acronyms used in this report.

AAFEBR	Anaerobic attached film expanded bed reactor
AF	Anaerobic filter
ABR	Anaerobic baffled reactor
AMBR	Anaerobic migrating blanket reactor
ANANOX	Anaerobic-anoxic-aerobic process
AnSBR	Anaerobic sequencing batch reactor
A/O	Anaerobic/Aerobic
AS	Activated sludge
BAF	Biological aerated filter
BOD	Biochemical Oxygen Demand
COD	Chemical Oxygen Demand
DHS	Downflow hanging sponge reactor
DO	Dissolved oxygen
DOC	Dissolved organic carbon
EBPR	Enhanced Biological Phosphorus Removal
HPH	High pressure homogenizer
HRT	Hydraulic retention time
JHB	Johannesberg EBPR process
LSP	Low sludge production process
MCAB	Membrane coupled anaerobic bioreactor
MLSS	Mixed liquor suspended solids
MLVSS	Mixed liquor volatile suspended solids
MUCT	Modified UCT process
nbVSS	non-biodegradable VSS
O	Aerobic system
OUR	Oxygen uptake rate
PAO	Phosphorus accumulating organisms
RAS	Return activated sludge
rbCOD	readily biodegradable COD
SBR	Sequencing batch reactor
SRT	Solids retention time
SVI	Sludge volume index
TCP	2,4,5-trichlorophenol
TCS	3,3',4',5-tetrachlorosalicylanilide
TSS	Total suspended solids
UASB	Upflow anaerobic sludge blanket reactor
UCT	University of Capetown EBPR process
VSS	Volatile suspended solids

EXECUTIVE SUMMARY

Aerobic biological processes used to treat wastewater produce a significantly higher quantity of sludge than anaerobic processes. This excess sludge requires further processing and disposal or recycle. For municipal wastewater treatment, typical amounts of excess sludge in the biological treatment process may range from 500–800 lbs. dry weight for each million gallons of wastewater treated. The handling and disposal of waste sludge represents a significant portion (40–50%) of a plant's operating cost. For industrial wastewaters, sludge processing may be more problematic because fewer options are available for the ultimate disposal/reuse of processed biosolids for these wastewaters.

The main goal of this research project was to evaluate technology that could minimize biomass production in the biological liquid-treatment processes, thereby reducing the total quantity of solids for processing and ultimate disposal/reuse. Reduced biomass production can lower total costs for sludge thickening, digestion, dewatering, hauling, and tipping fees. The first phase of this study was to identify and document methods reported to minimize biomass production in aerobic biological activated sludge processes. The technologies were evaluated in terms of their sludge reduction potential, ease of implementation into existing facilities, impacts on plant operations and effluent quality, reliability and risk associated with the processes, and relative capital and operating costs. Following the technology identification, a project workshop was held to evaluate the more promising technologies and to guide research for further understanding of these technologies. Two technologies were selected for further evaluation. One was the anaerobic-aerobic process used for enhanced biological phosphorous removal (EBPR), in which a chemical oxygen demand (COD) loss has been previously reported in the anaerobic step; a COD loss would result in lower biomass production and lower oxygen requirements. A number of bench-scale studies were carried out to evaluate the COD loss mechanisms for the anaerobic-aerobic EBPR process and to compare the amount of sludge reduction relative to an aerobic system only.

The second technology that showed promise for biomass reduction and ease of implementation was ozonation of return activated sludge (RAS) solids in the activated sludge process. Pilot plant and full scale testing indicated that the ozonation process, treating a portion of RAS, was feasible and a very effective method of biomass reduction. A detailed economic analysis was done on the ozonation process to compare the cost of implementing the process to the cost savings realized by reduced costs for sludge processing, hauling, and ultimate disposal.

Anaerobic treatment processes can result in much less biomass production compared to aerobic biological treatment processes, but the ability to produce a secondary quality effluent has not been shown with conventional anaerobic treatment processes. An anaerobic baffled reactor process followed by membrane separation has been proposed as a means to achieve a high quality effluent, as well as low biomass production. This process is described in Appendix A, along with a description of an ongoing pilot plant study. The performance of the pilot plant testing will be provided in a later report.

The following summary provides a brief description of the biomass reduction technologies evaluated, the range of biomass reduction efficiencies reported, and other significant findings from an assessment of the technologies.

ES.1 Anaerobic - Aerobic EBPR Process

Anaerobic/Aerobic (A/O) processes have been used in various activated sludge configurations for enhanced biological phosphorous removal (EBPR). The process involves placing an anaerobic contact zone ahead of an anoxic or aerobic zone in an activated sludge treatment system. The anaerobic contact detention time is approximately 45–60 minutes and the zone is mixed to provide contact between mixed liquor and influent wastewater. Various design configurations for this process are described in the report. The process was a good candidate for biomass reduction due to the ease of implementation into existing facilities, and additional benefits of nutrient removal and improved sludge settling characteristics.

Previous researchers have reported a COD loss in the anaerobic contact zone of A/O processes, based on the results of COD mass balances with bench-scale testing data. The COD balance included influent and effluent COD, waste solids COD, COD satisfied by nitrate/nitrite reduction, and COD satisfied by measured oxygen uptake rates in the aerobic zones. The COD loss calculated from the mass balances was 20–25%, which could thus lead to a 20–25% reduction in biomass production. One explanation given for the COD loss mechanism relates to the fermentation process in the anaerobic contact reactor, namely that hydrogen or other products produced in the fermentation process escaped the anaerobic reactor. Side by side bench-scale reactors were operated at the King County West Point Wastewater Treatment Facility in Seattle, WA to assess this mechanism by adding a fermentable and non-fermentable substrate, respectively. The feed to the anaerobic contact zone for one reactor was spiked with acetate and the feed to the other with lactate, a fermentable substrate. No significant change in sludge production was observed between the two reactor operations. COD loss was not observed from a COD mass balance for these reactor operations. Further tests were carried out at the University of Washington Environmental Laboratories to compare solids production in parallel operations of aerobic treatment only and anaerobic/aerobic treatment. The feed to the experimental reactors were synthetic wastewater or municipal wastewater plus synthetic wastewater. In both cases, no significant reduction in volatile solids sludge production was found between the two systems with and without an anaerobic contact period.

The explanation for COD loss put forth by previous researchers was further evaluated based on fundamental biological reaction and reactor mass transfer considerations. At issue was whether a significant amount of hydrogen or methane could be produced in the anaerobic contact zone and be transferred from the reactor liquid to the overlying gas phase. Recent information on the volatile fatty acid composition of primary sludge fermenters used in EBPR processes aided the evaluation. The evaluation found that COD loss in the anaerobic zone due to hydrogen or methane production and stripping is highly unlikely. The limited reports of COD loss in other bench-scale experiments may have been an artifact of the experimental design associated with the relative shallow reactors and high surface exposure. In addition, the previous reports did not provide any proof of a COD loss mechanism. In view of the negative findings on COD loss from this work, the lack of definitive proof in previous work, and the risk associated with the assumption of COD loss in A/O systems, the assumption of a COD loss for all activated sludge A/O process designs is not recommended. The erroneous assumption of a COD loss for an activated sludge facility can lead to it being under- designed in terms of oxygen requirements and sludge production. A more reliable design would be based on an assumption of no COD loss in the anaerobic contact zone until the specific operating conditions and mechanism for COD loss are proven.

ES.2 Anaerobic Processes

The ability of anaerobic treatment processes to reduce biomass production by 60–80% compared to aerobic treatment is well-established. Anaerobic treatment processes have been used in full-scale industrial wastewater treatment for relatively high strength wastewaters. The purpose of this evaluation was to assess experience with anaerobic treatment processes for the treatment of weaker strength industrial wastewaters and municipal wastewaters. Of special concern was the ability to provide a secondary treatment effluent quality at lower temperatures of 10–20 ^{0}C. A number of different anaerobic treatment processes have been shown capable of producing a secondary effluent when treating municipal wastewaters at the low temperature range. However, these studies were done mainly with small bench-scale or pilot-scale systems without long-term evaluations that included variable loadings. These processes included upflow-expanded bed, upflow anaerobic sludge blanket, anaerobic baffled reactor, and anaerobic treatment with membrane separation processes. The major concern for these processes is the ability to maintain high effluent solids capture efficiency and long solids retention times (SRTs) in the system to maximize the efficiency of methanogens and process stability. Further research is needed in larger scale systems with long-term operation to demonstrate the feasibility of this promising technology for biomass reduction and lower energy requirements. Conceivably the process can be applied to existing secondary treatment facilities and the retrofit cost and design will be site-specific. Besides a major reduction in sludge production, the process can provide a significant energy savings by avoiding aeration energy requirements for aerobic processes and producing methane from degradable organic material in the wastewater.

ES.3 Sequential Anaerobic - Aerobic Processes

The sequential anaerobic/aerobic process uses an aerobic step for final polishing after the anaerobic process to produce a secondary treatment quality effluent. A number of bench-scale and small pilot-scale studies have demonstrated the effectiveness of this process, treating domestic wastewaters with an overall biomass reduction of 40–60% compared to conventional aerobic processes. The anaerobic process accomplished 75–85% biochemical oxygen demand (BOD) removal prior to the aerobic process. This process was proven quite feasible for wastewater temperatures above 20 ^{0}C, but long-term larger pilot-scale testing is needed to demonstrate its reliability and feasibility. The process has yet to be evaluated for long-term operation at lower temperatures. Further research is needed to fully exploit this effective biomass reduction technology for low temperature applications.

ES.4 Cell Disruption by Ozonation

In this process a portion of the return activated sludge (RAS) flow is diverted to an ozone contactor where ozone is used to disrupt the bacteria cells and release soluble substrates for further degradation in the aeration tank. The ozone demand is in the range of 0.2–0.4 g O_3/g TSS destroyed. Biomass reductions in the range of 50-90% have been demonstrated in bench-scale, pilot-plant scale, and full-scale testing. Though the process was shown to be technically feasible and reliable, an extensive cost analysis showed that it was only economically attractive for cases where the sludge processing and disposal costs were unusually high (e.g., in cases where sludge incineration is needed or hauling distances are in the range of hundreds of mile).

ES.5 Cell Disruption by Mechanical Methods

Similar to the ozonation process, a portion of the RAS stream is treated in mechanical disintegration processes that can cause bacteria cell disruption and release of intracellular material. The devices used for mechanical disintegration may include high-speed homogenizers, ball mills, and high-pressure nozzles, all of which are energy intensive. One process, termed the KADY Biolysis process has been tested in bench and full-scale facilities and was evaluated in this study. No verifiable side-by-side studies with and without the disintegration process were available in which the activated sludge processes were operated at identical conditions. Full-scale plant testing of parallel activated sludge trains at the Springfield, MA facility, with and without the KADY process, showed 12% less sludge production for the KADY process, but the KADY process train was operated at a longer SRT. Such results suggest that if less sludge production is possible, it is less than 12%. The increased energy cost, operational complexity, and maintenance requirements may outweigh the modest biomass reduction potential for this process.

ES.6 Cell Disruption by Thermal Methods

In this process a portion of the RAS stream is contacted for 24 hours at high temperatures of 80–100 ^{0}C to cause cell lysis and release of intercellular components. In one version of this process, thermophilic bacteria were added to the thermal treatment reactor to enhance the cell degradation efficiency. Biomass reductions in the range of 20–80% have been shown. The high energy cost and space requirements associated with this process make it less attractive compared to other biomass reduction technologies.

ES.7 Chemical Uncoupling

Chemicals may be added to biological processes to interfere with the energy capture from biological oxidation processes. By making the energy produced from biological oxidation unavailable to the bacteria cell, there is insufficient energy for new cell growth and little or no sludge production occurs. Chemical uncoupling occurs when membrane-active chemicals, such as protonophores, cause leakage of the proton gradient across the cell membrane. The proton gradient is responsible for ATP generation. A large number of chemicals—including dinitrophenols and chlorinated phenols—have been used successfully to cause chemical uncoupling. Reductions in biomass yield in the range of 50–60% have been demonstrated in bench-scale tests. The major draw back for this process is the presence of toxic chemicals in the wastewater treatment plant effluent. Until that problem can be solved, this biomass reduction method will not be feasible.

ES.8 Predator Enhancement Process

The predator enhancement process relies on the consumption of dispersed bacteria by protozoa, which have a much lower biomass yield compared to bacteria. The process requires a separate first stage, high growth rate reactor (SRT of less than 12 hours), in which the rapid consumption of soluble substrates leads to a highly dispersed young bacteria cell culture that has not developed sufficient extracellular polysaccharide for good flocculation. Protozoa are then able to consume the highly dispersed bacteria cells. The high rate bioreactor is followed by a conventional activated sludge or fixed film process for final treatment. The process has been demonstrated with pulp and paper mill wastewaters, and an average biomass reduction of 50% has been shown. The process was less successful for municipal wastewater treatment, where the soluble BOD content is much lower compared to that for industrial wastewaters. The process is relatively easy to implement, but requires increased operator skill and monitoring to manage the more complex biological process. It is less compatible with biological nutrient removal, but this is of less importance for industrial wastewater applications, where nutrients often are limited.

ES.9 Extreme SRT Process

The Extreme SRT Process is an activated sludge system with SRTs greater than 50 days that can accomplish significant biomass reduction, presumably by long-term endogenous respiration. However, results of previous research on this process suggest that other mechanisms, such as predator activity, may play a role in the biomass reduction. Biomass reductions in the range of 70–90% have been demonstrated. With the development and application of membrane bioreactors for activated sludge treatment, extreme SRT processes are feasible for full-scale wastewater treatment. The large aeration tanks required for this process compared to conventional activated sludge treatment, with or without membrane separation, make the process less feasible for most full-scale applications. However, for small on-site treatment systems, where sludge minimization is desired, the process may have some potential.

The report includes a section (Chapter 12.0) summarizing various research needs for the different biomass reduction technologies. The report also addresses a number of factors related to the implementation of biomass reduction technologies, including operational effects, the ability to accomplish biological nutrient removal, and effects on plant performance and operating costs.

CHAPTER 1.0

INTRODUCTION

1.1 Project Goals

Aerobic biological treatment processes produce a significant quantity of excess sludge that requires further processing and disposal or recycling. Aerobic bacteria exhibit highly efficient organic substrate metabolism with the ability to incorporate 40–50% of the substrate, as chemical oxygen demand (COD), into cell mass. For municipal wastewater treatment, typical amounts of excess sludge production from conventional activated sludge processes may range from 500 to 800 lbs dry weight for each million gallons of wastewater treated (Crites and Tchobanoglous, 1998). The handling and disposal of the sludge represents a large fraction of plant operating costs. Problems associated with sludge production are just as significant for industrial wastewater treatment, and in some cases can be more problematic, where fewer options are available for the ultimate fate of the processed biosolids. Where the sludge is not processed on-site, but is hauled to sanitary landfills or to compost operations, the ability to reduce the sludge quantity can result in a significant cost savings. When treated on-site, cost savings also can be realized by reduced cost for dewatering and for tanks and equipment for sludge processing.

The focus of this research project was technology that could result in less biomass and total solids production in the liquid treatment process, and methods for sludge digestion or waste sludge processing were not considered. In all cases the ability to use identified biomass reduction technologies in retrofit applications was important, considering the number of full-scale facilities with existing liquid and solids processing systems. Advantages for biomass reduction in liquid processing are as follows:

1. The activated sludge aeration tank treatment capacity (i.e., the mass of biological oxygen demand–BOD–that can be treated daily) can be increased, provided that sufficient aeration and clarification capacity exist. For a given applied BOD load (Q x BOD), the aeration volume required is a function of the operating SRT, the net solids yield, and mixed liquor suspended solids (MLSS) concentration that can be maintained, as shown for a completely mixed activated sludge process. For constant MLSS and influent BOD concentrations and aeration volume in Equation 1.1-1, a reduction in the net solids yield due to applying a biomass reduction technology allows an increase in the influent flow

rate to the aeration tank. Thus, the plant treatment capacity can be increased provided there is sufficient hydraulic and oxygen transfer capacity.

$$V = Y_N \frac{(Q)\,(BODr)\,(SRT)}{MLSS} \qquad 1.1\text{-}1$$

Where:

V = volume, m^3

Y_N = net solids yield, g TSS/g BOD

Q = influent flow rate, m^3/d

BOD_r = BOD removed, mg/L

SRT = solids retention time, days

MLSS = aeration tank mixed liquor suspended solids conc., mg/L

2. The cost of solids processing, including electrical and chemical costs, will be decreased as these are normally in proportion to the amount of solids processed. The solids processing system would also have the ability to handle solids for treatment of a higher influent flow rate.
3. Ultimate disposal costs will be decreased, as these are normally proportional to the wet tons of solids hauled and disposed or reused.

The main goals of this research were to (1) identify and document a comprehensive review of methods to minimize biomass production in aerobic biological activated sludge processes, (2) evaluate the technologies and compare them with regard to sludge reduction potential, ease of implementation, plant impacts, risks, and operating costs,(3) carry out specific research projects to evaluate highly ranked technologies following a workshop with the project team and a review with the project steering committee, and (4) identify further research needs to develop promising technologies that produce less biomass from biotreatment processes.

1.2 Solids Production in Biological Wastewater Treatment

Biological wastewater treatment has proven to be the most cost effective method to remove organic pollutants from wastewater. In these processes, bacteria oxidize a portion of the organic substances to CO_2 and H_2O and use the energy from the oxidation reaction to produce additional biomass. The process also removes non-biodegradable organic solids and inert inorganic solids from the wastewater as part of its purpose to produce an effluent low in BOD and suspended solids. Thus, besides the treated liquid as a process output, it also produces a solids residual stream consisting of biomass, non-biodegradable volatile suspended solids (nbVSS), and inert inorganic material, that must be processed for beneficial reuse or ultimate disposal.

The solids production from biological treatment will vary from site to site depending on the wastewater characteristics and degree of treatment before the biological process. The following equation describes the solids production in activated sludge treatment as a function of the wastewater characteristics (Tchobanoglous et al., 2003):

$$P_x = \underbrace{\frac{Y(\Delta bCOD)Q}{[1+K_d(SRT)](0.85)}}_{(A)} + \underbrace{\frac{f_d K_d Y(\Delta bCOD)Q(SRT)}{[1+K_d(SRT)](0.85)}}_{(B)} + \underbrace{nbVSS(Q)}_{(C)} + \underbrace{X_I(Q)}_{(D)} \qquad 1.2\text{-}1$$

Where:

Px	= solids production, g TSS/d
A	= biomass component, g VSS/d
B	= biomass cell debris component from endogenous decay, g VSS/d
C	= non-biodegradable VSS component, g VSS/d
D	= influent inorganic solids component, g TSS/d
Q	= flow, m^3/d
0.85	= g VSS/g TSS for biomass
Y	= biomass synthesis yield, g VSS/g bCOD removed
ΔbCOD	= biodegradable COD removed, mg/L
K_d	= endogenous decay coefficient, gVSS/gVSS–d
f_d	= cell debris production coefficient, g VSS debris/g VSS cells
SRT	= solids retention time, days
nbVSS	= influent non-biodegradable VSS, mg/L
X_I	= influent inorganic TSS concentration, mg/L

Table 1 provides typical values used for the coefficients in Equation 1.2-1.

Table 1-1. Activated Sludge Kinetic Coefficients at 20 °C (Tchobanoglous et al., 2003).

Coefficient	Unit	Typical Value
Y	g VSS/g bCOD	0.40
K_d	g VSS/g VSS –d	0.08
f_d	g VSS/ g VSS	0.15

The value for K_d is corrected for temperature as follows:

$$K_{d,T} = K_{d,20} (\theta)^{T-20} \qquad 1.2\text{-}2$$

Where:

$K_{d,T}$ = endogenous decay coefficient at temperature T.

T = temperature, °C

θ = temperature correction coefficient

A value of 1.04 has been used for θ in Equation 1.2-2 (Tchobanoglous et al., 2003).

The influent wastewater nbVSS concentration has a significant impact on the solids production rate, as well as the activated sludge process SRT. For domestic wastewaters the nbVSS concentration for primary treated effluents may be in the range of 15–30 mg/L, and with no primary treatment it may be in the range of 60–100 mg/L (Tchobanoglous et al., 2003). Figure 1-1 illustrates the effect of SRT on the observed solids production in g VSS/g COD removed for a municipal wastewater with and without primary treatment, and for an industrial wastewater consisting mainly of soluble fully biodegradable organics in which the only solids production is from biomass. For this example the coefficients in Table 1 were used. Assumptions used are nbVSS concentrations of 25 mg/L and 80 mg/L, with and without primary treatment, respectively, and an influent biodegradable COD of 200 mg/L and 320 mg/L, with and without primary treatment, respectively. The observed yields calculated for a 5-day SRT, for example, are 0.43, 0.37, and 0.31 g VSS/g CODr for domestic wastewater with primary treatment, domestic wastewater without primary treatment, and a totally soluble biodegradable wastewater, respectively. For the domestic wastewater the biomass portion represents 72-83% of the solids yield at the 5-day SRT, but at longer SRTs it is a lower fraction. At a 20-day SRT, the biomass fraction is about 55-70% of the solids produced for the domestic wastewater without and with primary treatment, respectively. This example shows that technologies that can reduce biomass production can significantly affect the total solids production, and that the effect may be more pronounced for industrial wastewater containing little nbVSS.

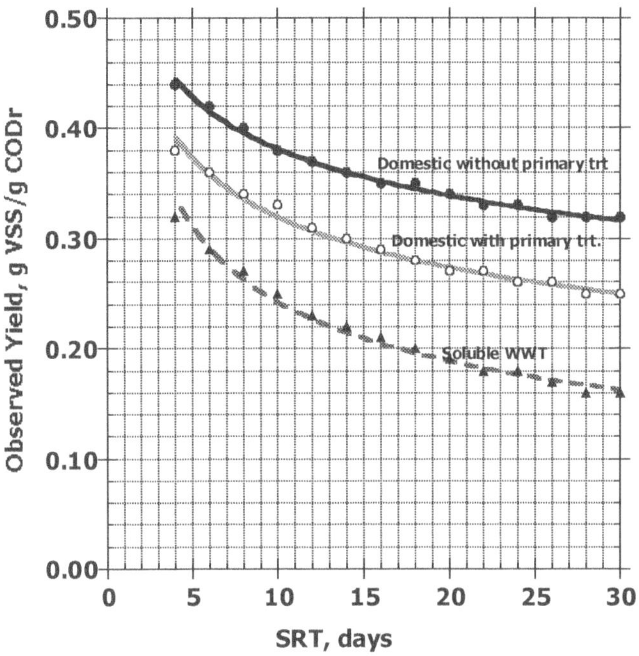

Figure 1-1. Example of Observed Solids Yields as a Function of SRT for Domestic Wastewater With and Without Primary Treatment, and for a Soluble Biodegradable Industrial Wastewater.

The evaluation of solids yield given in Figure 1-1 provides a background for evaluating observed solids production reported for the technologies evaluated in this report. Assuming a ratio of mixed liquor volatile suspended solids (MLVSS) to MLSS of 0.80 for domestic wastewater treatment and a bCOD/BOD ratio of 1.6 (Tchobanoglous et al., 2003), typical solids yields on the basis of COD or BOD removed and VSS or TSS are summarized in Table 1-2 for domestic wastewater treatment.

Solids yields and alternative electron acceptors oxidation-reduction reactions by bacteria provide the energy needed for cell synthesis. A given amount of energy is needed for cell production, which is provided by the oxidation-reduction reactions. For heterotrophic bacteria, biodegradable

Table 1-2. Typical Observed Solids Yield for Activated Sludge Treatment of Domestic Wastewater (Data from Tchobanoglous et al., 2003).

Yield Parameter	Value
g VSS/ g COD	0.40
g TSS/ g COD	0.50
g VSS/ g BOD	0.65
g TSS/ g BOD	0.80

organic compounds are oxidized, resulting in the transfer of electrons from a donor to an ultimate electron acceptor to complete the reaction. Though oxygen is the normal electron acceptor for activated sludge treatment, nitrate also may provide an electron acceptor for removal of a portion of the influent BOD in systems designed for nitrification and denitrification. Because there is less energy produced when nitrate serves as the electron acceptor instead of oxygen, more of the organic substrate must be oxidized to produce enough energy for a unit of cell growth. The result is that a smaller proportion of the substrate is then available for cell production, which results in a lower synthesis yield in terms of g VSS/g COD removed. Table 1-3 compares synthesis yield values for aerobic, anoxic, and anaerobic respiration.

Table 1-3. Typical Synthesis Yields for Different Electron Acceptors for Biological Oxidation of Organic Material (Data from Tchobanoglous et al., 2003).

Condition	Electron Acceptor	Synthesis Yield g VSS/ g COD	Products
Aerobic	Oxygen	0.40	CO_2, H_2O
Anoxic	Nitrate	0.30	N_2, CO_2, H_2O
Anaerobic	Organics	0.06	Organic acids
Anaerobic	CO_2	0.05	Methane

For respiration with nitrate, the biomass production is lower by about 25% due to less energy available with nitrate as the electron acceptor. McClintock et al. (1988) and Mines and Kalch (1999) reported a lower biomass production for nitrate respiration compared to aerobic respiration by 20–25%. For a typical domestic wastewater primary effluent with 120 mg/L BOD and 25 mg/L of oxidizable nitrogen, about half of the BOD can be oxidized by nitrate produced in an anoxic/aerobic process. Based on a 25% lower synthesis solids yield with nitrate use in lieu of oxygen, this would theoretically result in a reduction in biomass production of about 12%.

For anaerobic processes the biomass production is considerably less compared to aerobic processes. Based on the yield values shown in Table1-3, the biomass production for an anaerobic process may be about 25% of that for an aerobic process. In terms of the magnitude of biomass reduction, this was the most attractive technology in meeting the goals of this study. Various anaerobic process designs have been applied successfully for treatment of high strength industrial wastewaters and at warm temperatures. In Chapter 2.0, anaerobic processes are described and their reported performance evaluated for treatment of both higher and lower strength wastewaters, such as domestic wastewaters. The literature review also pursued anaerobic applications at lower temperatures (less than 20 ^{0}C).

1.3 Technologies Reported to Reduce Biomass Production

A number of sources—including literature, vendors, other researchers, and practicing engineers—were consulted to identify technologies that reported reduced biomass production. The technologies selected for further evaluation were those for which a mechanism for biomass reduction could be defined, laboratory, pilot, or full-scale performance data were available, and initial information suggested biomass reduction existed. The technologies evaluated are listed as follows, in the order of their presentation in this report.

- ***Anaerobic Processes*** that rely solely on biological anaerobic reactors for wastewater treatment.

- ***Sequential Anaerobic-aerobic Treatment Processes*** that use anaerobic treatment as a preliminary treatment step before an aerobic process, which is added primarily to meet secondary treatment performance levels.

- ***Anaerobic-aerobic Enhanced Biological Phosphorus Removal Processes (EBPR)*** that incorporate a reactor for anaerobic contacting of mixed liquor and influent wastewater before subsequent aerobic treatment steps, with the primary goal of providing phosphorus removal by biological uptake and storage.

- ***Cell Disruption Processes*** that disrupt cell integrity by chemical or mechanical means to cause cell lysis and release of cellular contents.

- ***Chemical Uncoupling Processes*** that use chemicals that interfere with the capture and utilization of energy by cells during oxidative respiration.

- ***Predator Enhancement Processes*** that use higher life forms, with lower growth yields, to consume bacteria that have assimilated the wastewater degradable organic material.

- ***Extreme Solids Retention Time (SRT) Processes*** that use unconventional design methods to provide excessive SRTs to maximize endogenous decay and minimize net biomass production.

1.4 Project Workshop and Technology Prioritization for Experimental Phase

The above technologies were evaluated and ranked by the project technical team during the first year to identify priorities for further bench-scale or full-scale experimentation. In addition to those shown, a process termed Bioconversion was also evaluated. An example of this process is the Micronaire process (Bizier, 1999) that processes waste sludge in a side reactor by classification and fine bubble flotation. Because this process is not used with liquid treatment, it was not considered further after the process evaluation workshop. The results of the technical team workshop were reviewed with the project steering committee to develop a final prioritization and selection of processes for the experimental phase.

The technology evaluation criteria used for the workshop are shown in Table 1-4, along with the weighting factors developed by consensus amongst the workshop participants. The main areas of concern were reliability and proof of concept, implementation impacts, and cost-effectiveness. Biomass reduction level, costs, impact on plant performance, and impact on plant operations and maintenance were key factors. Costs were not developed in detail but were considered on their relative magnitude for implementation of the biomass reduction technology at an existing facility. The decision analysis tables and accumulative points scored by the workshop participants are summarized in Appendix A.

Table 1-4. Biomass Reduction Technology Evaluation Criteria.

Criteria	Weight Factor
I. Certainty/Reliability	
Level of Demonstration	3
Quality of Evidence	6
Process Stability	5
II. Implementation Impacts	
Biomass Reduction	10
Ease of Implementation	3
Impact on Operations/Maintenance	7
Impact on Plant Performance	8
III. Cost Effectiveness	
Sludge Treatment & Disposal Cost	10
Plant Capacity Impacts	4

Following a review workshop with the Water Environment Research Foundation (WERF) project steering committee, the biomass reduction technologies were ranked as shown below. The processes that received higher rankings also had a higher priority for the experimental phase.

1. Anaerobic-aerobic EBPR processes
2. Anaerobic processes
3. Cell disruption by ozonation
4. Extreme SRT processes
5. Cell disruption by mechanical methods
6. Chemical uncoupling processes
7. Predator enhancement processes

Research goals were identified to test the biomass reduction and feasibility of the processes, with the intent to evaluate at least two processes experimentally within the project resources. Testing and evaluation were to be done by project contributors. For the EBPR process, biomass reduction was predicated on a COD loss mechanism claimed by previous researchers. A series of bench-scale tests were done to evaluate the feasibility of this biomass reduction potential and the results are summarized in Chapter 10.0. Full-scale tests were planned but did not occur due to an unexpected plant aeration replacement program. A pilot plant facility was built to evaluate the feasibility of an anaerobic baffled reactor followed by membrane separation for the treatment of low strength, low temperature wastewaters. The test will be ongoing at the King County West Point wastewater treatment plant beyond this project time table. An economic analysis was done on the ozonation cell disruption technology as part of the feasibility evaluation of that cell disruption technology, and the results are presented in Chapter 6.0.

The evaluations of the various technologies are presented in Chapters 2.0 through 9.0. Each chapter includes a description of the process, the fundamental biomass reduction mechanism, reported process performance, plant retrofit considerations, overall assessment, and research needs.

CHAPTER 2.0

ANAEROBIC PROCESSES

2.1 Description of Anaerobic Wastewater Treatment Processes

Anaerobic degradation is one of the oldest technologies for stabilizing organics in wastewaters. Septic systems, Imhoff tanks and other anaerobic methods rivaled aerobic processes early in the twentieth century. Aerobic processes have surpassed anaerobic treatment for treating liquid wastewater streams, except in the case of high organic concentration industrial wastewaters, due to the ability to produce a higher quality effluent, compactness, reliability and ease of operation. However, research on anaerobic methods has continued to the present time and anaerobic wastewater treatment processes are showing promise for treatment of moderate strength wastewaters at moderate temperatures.

Many reactor configurations have been proposed for anaerobic biotreatment of wastewater. Recent research has focused on reactor designs that provide sufficient retention of biomass to sustain anaerobic biological activity even at low temperatures and with relatively dilute wastewaters.

2.1.1 Anaerobic Biofilm Reactors

Biofilm reactor configurations maintain a high reactor biomass concentration due the growth and maintenance of the biomass on a support media. The most frequently studied reactor configurations for anaerobic treatment of wastewaters are the anaerobic attached film expanded bed reactor (AAFEBR) (Jewell et al., 1981; Alderman et al., 1998; Collins et al., 1998) and the anaerobic filter (AF) (Bodik et al., 2002a; Elmitwalli et al., 2002a). The AAFEBR consists of an upflow suspended bed of a medium, such as sand or activated carbon, on which the anaerobic biofilm grows. A recirculation pump recycles reactor effluent to maintain a sufficient upflow velocity for expansion of the bed (Figure 2-1). For silica sand media with a diameter in the range of 0.2–0.5 mm and specific gravity of 2.65, an upflow velocity of about 2 m/h (0.11 ft/min) is used. In the anaerobic filter (AF) the media is stationary, the wastewater flow is either upflow or downflow, and no recirculation is usually used except for high-strength wastewaters. While the early applications contained rock, a variety of designs employing synthetic plastic packing (pall rings, random plastic, tubular plastic, and plastic crossflow media) are more common today. In upflow applications a low upflow velocity is generally used to prevent washing out of the biomass. For downflow reactors a packing with a higher void volume is generally used to prevent plugging problems. Over time, solids and biomass will accumulate in the packing, causing plugging and flow short-circuiting. At this point excess solids must be removed by flushing (backwashing) and draining the packing (Tchobanoglous et al., 2003).

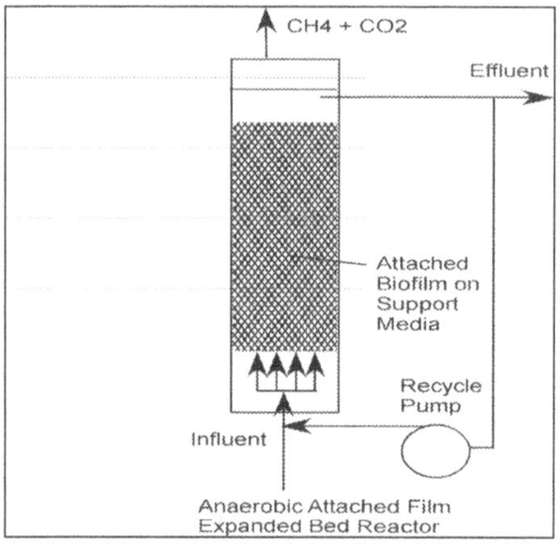

Figure 2-1. AAFEBR Reactor Schematic.

2.1.2 Staged Anaerobic Reactors

Two types of staged anaerobic reactors have been developed in the past few years, the anaerobic baffled reactor (ABR) and the anaerobic migrating blanket reactor (AMBR). In the ABR process, the baffled reactor is divided into three or four stages. Baffles ahead of each stage are used to direct the flow downward from the previous stage and then in an upflow mode through the sludge in each stage. A number of modifications have been used to maintain and capture the solids in each of the baffled reactor stages (Barber and Stuckey, 1999). Though granulated sludge is not considered essential for the operation of the ABR process, it has been observed in the ABR systems (Boopathy and Tilche, 1992). Advantages claimed for the ABR process include: (1) simplicity with no packing material or mechanical mixing, (2) a long SRT relative to the hydraulic retention time, (3) a staged operation to improve kinetics, and (4) a stable operation under shock loads.

The AMBR process is similar to the ABR with the added features of mechanical mixing in each stage and an operating approach to maintain the sludge in the system without using packing or settlers for additional solids capture. In the AMBR system, the sludge migrates from the inlet feed stage to the latter stages, as the wastewater travels through the reactor. The influent feed point is changed periodically to the effluent side and the effluent withdrawal point also is changed to the opposite end. In this way the sludge blanket remains more uniform in the anaerobic reactor (Angenent et al., 2001; Angenent and Sung, 2001).

Plug flow type of processing also is achieved in the anaerobic sequencing batch reactor (AnSRB) (Ndon and Dague, 1997a; Dague et al., 1998; Bagley and Brodkorb, 1999; Bodik et al.,

2002a). In this system the mixed, enclosed anaerobic reactor is batch fed, followed by react, settle, and effluent withdrawal steps.

2.1.3 Upflow Anaerobic Sludge Blanket Reactors

Upflow anaerobic sludge blanket reactors (UASB) provide anaerobic treatment in an upflow system with a high biomass and long SRT. Most UASB systems involve a highly flocculent or granular biomass suspension. The first applications of the upflow sludge blanket technique and identification of the importance of the granular biomass formation was by Lettinga and coworkers in the Netherlands (Lettinga et al., 1980). In the UASB process, wastewater is distributed at the bottom of the reactor and travels in an upflow mode through a sludge blanket at velocity of about 1 m/h (Figure 2-2). Solids separation and gas collection occurs at the top of the reactor. A key feature of the process is the development of a dense granulated sludge with solids concentrations of 50–100 g/L at the bottom of the reactor and 5–40 g/L in the more diffused upper zone. The formation of the granulated sludge is affected by the wastewater composition (high carbohydrate wastes favor granulation) and the presence of particulate organics (Tchobanoglous et al., 2003).

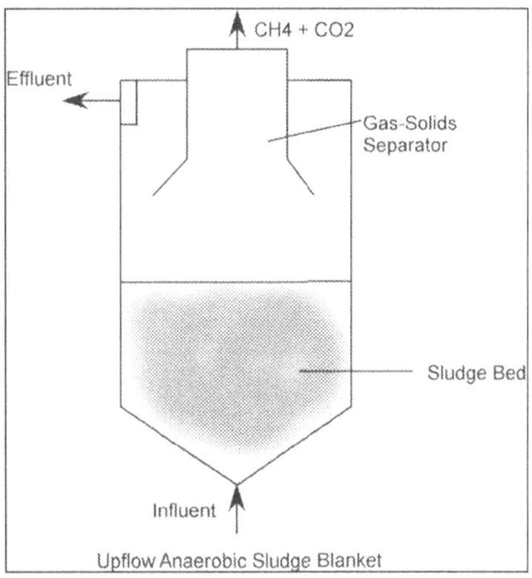

Figure 2-2. UASB Bioreactor Schematic.

2.1.4 Membrane Coupled Anaerobic Bioreactors

Retention of biomass in anaerobic bioreactors by recirculation of the bioreactor contents through crossflow membrane microfilter units was proposed two decades ago. The combination of anaerobic bioreactors with ultrafiltration membrane separation of the biomass from the wastewater solution provides an ultimate retention of cells. Membrane coupled anaerobic bioreactors (MCAB) have been studied in various configurations, which combine the membrane separation unit either external to the bioreactor (Figure 2-3) or suspended within the reactor itself (Choo and Lee, 1996a, 1996b, 1998; Hernandez et al., 2002).

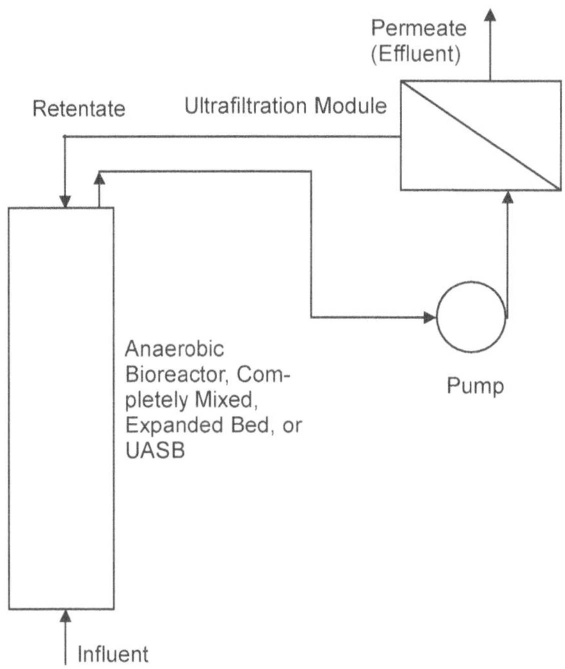

Figure 2-3. Schematic of Membrane Coupled Anaerobic Bioreactor with External Membrane Separation.

2.2 Process Fundamentals

The mechanism of biomass reduction for anaerobic bioprocesses, compared to aerobic processes, is related to the differences in electron acceptors and reaction products. In anaerobic processes the two main reaction steps, fermentation and methanogenesis, produce products (volatile fatty acids and methane, respectively) that are highly reduced compounds compared to the production of highly oxidized products (CO_2 and H_2O) for aerobic reactions. With less energy available per unit of substrate processed to support cell growth, the biomass yield is much lower. As given in Table 1-3, the biomass yield for the coupled reaction of fermentation and methanogenesis is 0.11 g VSS/g COD removed, compared to 0.40 g VSS/g COD for aerobic reactions. Indeed, the concept of lower biomass yields of anaerobic processes is so well accepted that most studies of anaerobic wastewater treatment processes have not reported solids production or cell yield. Usually sludge wasting was not practiced except through losses in the effluents. Thus, in the performance data summaries that follow little information is provided on solids yield because it was not reported, and in some cases no manual solids wasting was necessary, as the solids produced were lost in the effluent flow. The key issue with the application of anaerobic processes to reduce biomass production is their ability to produce an effluent quality comparable to aerobic processes and sufficient to meet secondary treatment discharge standards (85% BOD and TSS removal for municipal wastewaters). Thus, much of information gained from the review of anaerobic processes is aimed at treatment performance.

2.3 Reported Process Performance

The research team examined more than 50 studies reported in the literature for the effects of temperature, influent COD concentration, and organic loading rate on treatment efficiency. Table 2-1 lists the studies summarized here and the types of reactors involved, including UASB, AAFEBR, AF, ABR, AMBR, and AnSBR. Table 2-2 summarizes the research on MCAB reactors, including membrane characteristics.

Figure 2-4 shows the effects of temperature on process efficiency. Temperatures below 20 °C resulted in decreases in process COD removal for all process types. Nonetheless, several reactor types are able to meet an 85% COD removal requirement at temperatures of 12 °C or less, including AMBR, UASB, AAFEBR, ABR, and MCAB treatment schemes. Low temperatures apparently affect the MCAB process to a lesser extent (90% COD removal at 13 °C, as shown in Figure 2-4), presumably due to the forced retention of solids (Wen et al., 1999), although it should be noted that there are few studies of the MCAB process with low temperature, low concentrations wastewaters.

COD removal also fell for influent COD concentrations below about 400 mg COD/L (shown sorted by process type in Figure 2-5 and for each process separately in Figures 2-6 through Figure 2-13). One explanation for the decrease in total COD removal at lower COD concentrations is the role of decreased solids hydrolysis and removal, since many of these studies included particulate and colloidal solids in their influent. As total COD concentration fell, suspended and colloidal solids comprised a greater proportion of the organic load and the effluent COD. As expected, COD removals in MCAB reactors were unaffected by suspended solids in the effluent (open diamonds in Figure 2-5). Another explanation for poorer performance at lower influent COD concentrations is that a greater proportion of biomass produced is lost for operations with lower feed COD concentrations, and thus the reactors with lower COD feed concentrations have lower SRTs. Assuming that for most reactor applications, the solids separation method results in an effluent biomass concentration within a certain range, then a greater proportion of biomass produced from utilization of the influent COD will result in the loss of a larger fraction of the biomass as the wastewater is weaker. For higher feed concentrations, the higher amount of biomass produced results in a higher reactor solids concentration to help capture solids.

On the other hand, organic loading rate was not a significant determinant of process efficiency (Figures 2-16 through 2-23), although efficiencies tended to fall off at loading rates less than 10 g/L/d, perhaps due to the prevalence of low influent COD concentrations, a proportionately higher biomass loss, and proportionately higher particulate and colloidal solids in these studies.

Anaerobic processes were able to achieve effluent total COD concentrations of 50 mg/L or less treating municipal wastewaters at temperatures of 10 °C or less (Figure 2-4). Bodik et al. (2002a) determined the BOD5 and COD of an effluent of an AnSBR treating municipal wastewater, yielding a ratio of BOD/COD of 0.46. Applying this ratio to a typical regulatory limit of 30 mg BOD/L, an acceptable effluent COD for an anaerobic reactor treating municipal wastewater would be about 65 mg COD/L. From Figure 2-14, the anaerobic reactors that were able to achieve this effluent level at temperatures less than 12 °C were the UASB, ABR,

AAFEB, AF, and hybrid configurations. The UASB and AAFEB reactor configurations consistently turned out the lowest effluent COD levels at low temperature. However, it should be noted that some reactors, such as the MCAB, were not tested in the low temperature range.

Similarly, the AAFEBR, AF, and ABR configurations were able to meet a typical secondary treatment effluent requirement of 30 mg SS/L (Figure 2-15). Unfortunately, low temperature suspended solids removal for the UASB reactor was not reported in the studies available to the project team. By their nature, the MCAB treatment systems have virtually no suspended solids in their effluent. Figure 2-15 shows that the AnSBR system was unable to achieve 30 mg SS/L in the effluent for temperatures less than 15 °C.

Table 2-3 shows the operating characteristics for those anaerobic biotreatment systems that achieved effluents with CODs less than or equal to 65 mg/L treating municipal wastewaters at temperatures of 20 °C or less. The highest loading rate under these conditions was achieved by AAFEBR systems, which operated successfully with organic loading rates equal to 7.7–9.4 g COD/L/d in two studies. Systems that did not require pumped recycle streams were the ABR (organic loading rate, OLR, of 2.2 g/L/d) and UASB (4.1 g/L/d). The hybrid bioreactor study by Elmitwalli et al. (2002b) achieved a sufficiently low effluent COD, but effluent SS exceeded 30 mg/L. The single low temperature municipal wastewater study using an MCAB operated successfully with a loading rate of 4.2 g/L/d. It is important to note that all of the studies were conducted at bench scale.

As noted above, the solids yields for anaerobic processes usually are not reported. However, the yields for methanogenic processes typically range from 0.1 to 0.2 g biomass per g COD removed.

2.4 Plant Retrofit Considerations

Retrofitting requirements for anaerobic processes can be extensive, especially if membrane separation also is involved. Aeration tanks may be converted to anaerobic bioreactors, but specific site conditions will determine if the existing plant tank configurations are compatible with some anaerobic bioreactor schemes (for example, UASB or AAFEB). The membrane separation units may be used to replace existing clarifiers, which in some cases can be retrofitted to provide additional anaerobic bioreactor capacity.

2.5 Assessment of Process

Successful applications of anaerobic treatment to high strength, high temperature industrial wastewaters in the last 20 years have demonstrated the usefulness and efficiency of this treatment method. Several recent reviews summarize these applications (Macarie, 2000; Frankin, 2001; Lettinga, 2001; van Lier et al., 2001). Anaerobic wastewater treatment has come to be considered as the most cost-effective solution for industrial waste streams containing high levels of organics. Staged reactor systems have permitted the application of anaerobic treatment to toxic wastewaters and to extremes of temperature (van Lier et al., 2001). Anaerobic processes have also been used frequently as pretreatment processes for industrial wastewater systems (El-Gohary et al., 1999). UASB and AAFEB systems dominate industrial applications (Frankin, 2001).

As shown in Figures 2-14 and 2-15, several of the anaerobic systems are able to achieve acceptable secondary effluent quality treating municipal wastewaters at temperatures from 5–12 °C. Some of the wastewaters used in the studies are more concentrated than are typical of municipal wastewaters in the United States. More dilute wastewaters may pose problems in achieving suitable effluent quality.

There have been few studies of membrane processes applied to municipal wastewater treatment, but those available suggest that the MCAB process has desirable qualities for this application. The ability of the MCAB process to retain colloidal and particulate fractions results in especially low effluent COD and SS levels. The accumulation of these recalcitrant fractions in the bioreactor may allow greater time for their hydrolysis and digestion. Interestingly, methanogenic granulation was unaffected in a UASB reactor coupled to a membrane separator for treatment of a high strength molasses wastewater (Hernandez et al., 2002). Furthermore, coliform bacteria are removed in the MCAB process by six orders of magnitude when treating a domestic wastewater (Ueda and Hata, 1999).

The MCAB processes involve additional capital and operating costs. Membranes treating municipal wastewaters were able to operate at membrane fluxes ranging from 5–20 $L/m^2/h$, with transmembrane pressures of 4–70 kPa. The submerged membrane bioreactor used by Ueda and Hata (Ueda et al. 1999) was particularly economical, consisting of a membrane module submerged under a pressure head which ranged from 2 to 17 kPa. The membrane module tank was aerated, with sludge recirculated between it and the anaerobic tank, promoting nitrogen removal via nitrification and denitrification, while maintaining turbulent flow over the membranes which helped to minimize cake formation. The results suggested that membrane washing was required only once a year. Power consumption was 2.4 kWh/m^3 of treated water.

2.6 Research Needs

The primary issues for the practical application of anaerobic wastewater treatment processes are:

- Efficient retention of high concentrations of biomass in the reactor;
- Process activity at low temperatures typical of wastewaters in temperate climates;
- Sufficient retention time for hydrolysis of solids at low temperatures;
- Sufficient pathogen removal;
- Possible requirement for alkalinity addition;
- Process stability in the face of organic and hydraulic shock loadings of toxicants;
- Odor;
- Loss of effectiveness due to sulfate reduction;
- Lengthy start-up times;
- Effluent polishing to achieve secondary treatment effluent quality;
- Lack of experience with low concentration wastewaters at low temperatures;
- Lack of experience with large-scale systems treating municipal wastewaters.

These issues have been addressed in one form or another in the extensive literature of anaerobic wastewater treatment.

Bench-scale investigations of various types of shock loadings have not revealed significant problems, especially with fixed film systems (Collins et al., 1998). Large organic loading shocks lead to buildup of organic acids, which may lead to process failure, but modern instrumentation allows such changes to be monitored in time for process control changes to be implemented. Odor problems are a significant factor influencing public acceptance of large-scale processes, but odor treatment technologies are available that can limit the impact of this issue. Sulfate interference with methane production will be an issue in application of anaerobic processes in some locales, but the experiments by Parkin et al. (1991) show that anaerobic processes can operate successfully with about 10% of the carbon oxidation going to sulfate reduction.

Lengthy start-ups have been avoided by large inoculations with established granular sludge, which can be successfully stored for years. Effluent polishing in aerobic processes is necessary to avoid odor problems and oxygen demand loads on receiving waters (especially in the case of sulfate-laden waters). Effluent polishing adds to total cost and produces significant solids that must be considered when evaluating total costs. Experience with large-scale municipal wastewater treatment in anaerobic processes exists in warmer climates, especially in South America (Chernicharo and Cardoso, 1999; Goncalves et al., 1999; von Sperling et al., 2001). This experience base can be used to guide applications in temperate climates.

All anaerobic processes investigated to date share the problem of nutrient removal. Enhance biological phosphorus removal (EBPR) is not compatible with existing systems. Chemical precipitation will probably be needed if phosphorus removal is required for an anaerobic wastewater treatment process, unless unproven biological processes such as phosphine generation/stripping can be demonstrated (Roels and Verstraete, 2001). The addition of chemicals and phosphorus precipitation would increase the amount of sludge production for the anaerobic process. For an aluminum phosphorus precipitate, about 4 mg of solids are produced per mg of phosphorus removed, which would be equal to about 32 mg/L of solids production for a typical municipal wastewater with an influent phosphorus concentration of 8.0 mg/L. This compares to an estimated biomass production of about 20 mg/L for an anaerobic process and 80 mg/L for an aerobic process. Thus, the anaerobic process solids production with phosphorus precipitation is still less than that for an aerobic treatment process, but the solids production is more than doubled. Nitrogen removal may be accomplished by recirculating effluent from a subsequent aerobic nitrification step or by integration with the anaerobic-anoxic-aerobic (ANANOX) process (Garuti et al., 1992; Tilche et al., 1996).

The more challenging problems for application of anaerobic methods to the treatment of low concentration, low temperature municipal wastewater streams are the accumulation of suspended solids and colloidal COD in the biomass and the insignificant removal of pathogens. One approach to overcoming this problem is to remove suspended solids and colloids prior to anaerobic treatment, either by precipitation using coagulants (Kalogo and Verstraete, 2000) or by a two-step process aimed at partial hydrolysis of the suspended solids. The two-step process has been developed by Lettinga's group with a focus on an AF for the hydrolysis stage designed to have minimal suspended growth in order to avoid clogging (Elmitwalli et al. 2002a; 2002b).

However, in order to provide for colloidal COD degradation the second stage UASB/AF hybrid reactor required a hydraulic retention time (HRT) of eight hours.

Solutions to the problem of pathogen destruction in anaerobic treatment methods that rely upon attached growth or gravitational settling as separation methods require the use of post-treatment disinfection, either chlorination, ozonation, or ultraviolet. Membrane separation has the potential for avoiding or minimizing post-treatment disinfection. Membrane separation, while expensive, has been successfully applied to aerobic biotreatment of wastewaters. It seems reasonable that coupling membrane separation with an anaerobic process will lead to very high biomass levels in the bioreactor due to the efficient retention of cells by the membranes. Unfortunately, experiments with membrane separation coupled to various anaerobic bioreactors have revealed problems associated with the excessive accumulation of filter cakes on the surface of the membranes (Choo and Lee, 1996b).

Flux of anaerobic digestion effluents through the membrane is characterized by a rapid decline in permeation flux as the cake layer is deposited on the membrane surface, followed by a gradual and irreversible decline in permeation flux that occurs over several weeks (Choo and Lee, 1996b). The irreversible decline is apparently due to the penetration of the polarized layer by single cells that bridge the remaining gap to the membrane surface due to the formation of extracellular polymeric substances. Formation of this irreversibly bound layer may be minimized by limiting the permeation flux, by recirculating only the supernatant of a sludge bed reactor, and by wasting specific particle fractions of the solids.

As shown in Table 2-2, recent experiments with MCABs have demonstrated operation for more than 75 d with fluxes of 0.18–0.32 L/m^2/h (Lee et al., 2001; Fuchs et al. 2003). Higher fluxes have not been sustainable, resulting in frequent membrane changes (Hernandez et al., 2002), precipitous decline in suspended biomass (Choo and Lee, 1996a, 1996b, 1998), or excessive transmembrane pressure (Wen et al., 1999). These issues will determine whether the use of MCAB is cost effective.

The ability of MCAB to prevent loss of suspended solids at low COD concentrations and low temperatures suggests that the unhydrolyzed suspended solids may accumulate in the bioreactor. The few studies that investigated MCAB operation under these conditions have not explored the fate of these solids. Do they accumulate or is hydrolysis promoted by higher concentrations in the bioreactor? The combination of the UASB and ultrafiltration separation of suspended solids from wastewater apparently does not disrupt the formed granules in the UASB process (Lee et al., 2001; Hernandez et al., 2002). Further research is needed on the performance of MCABs with low influent COD concentrations, low temperature wastewaters.

Table 2-1. Anaerobic Bioreactor Studies.

Reactor type	Volume, L	Wastewater Type	Duration, d	Influent COD Concentration, mg/L	Temp, °C	Organic Loading, kg COD m^{-3} d^{-1}	COD Removal Efficiency, %	Comments	Reference
AAFEBR	2	Primary municipal	30 x 4	135-270	5-20	0.34-4.4	35-77	Model and cost analysis	Alderman et al., 1998
AAFEBR	4.3	Soluble synthetic	220	600-800	10	7-12	60-90		Rebac, 1998
AAFEBR	—	Soluble synthetic	—	200-600	10	4-16	40-80		Switzenbaum and Jewell, 1978 cited in Rebac, 1998
AAFEBR	—	Soluble synthetic	—	2600	12	2	50		De Man et al., 1988b cited in Rebac, 1998
AAFEBR	—	Municipal	—	760	10	8.9	53-85		Sanz and Fdzpolanco, 1990 cited in Rebac, 1998
AAFEBR	—	Municipal	—	300	9-11	4.5	20-48		Vanderlast and Lettinga 1992 cited in Rebac, 1998
AAFEBR	4.2	Synthetic	120-180	5000	35	14	95-98		Tay and Zhang, 2000
AAFEBR		Primary municipal		391	13-15	4.7-9.4	91-96		Vanderlast and Lettinga, 1992 cited in Elmitwalli et al., 2001
AAFEBR	1	Primary municipal	200	80-225	20	0.7-6.7	2-85		Jewell et al., 1981
AAFEBR		Synthetic		557-700	13-31	11.2-13.4	62-71		Maragno and Campos, 1992 cited in Ndon and Dague, 1997b
ABR	14.7	Synthetic	5-80		15-27	0.02-0.66	84-92		Manariotis and Grigoropoulos, 2002
ABR	10	Milk, Colloidal	2-75	500	35	0.15-9.2	25-98	Soluble microbial products increased with OLR.	Langenhoff et al., 2000a
ABR	10	Milk synthetic	30-70	500	10-35	1.2	63-90	No T-dependent selection	Langenhoff and Stuckey, 2000b
ABR	6.3	Carbohydrate-protein, Synthetic gray water		480-7600	29-35	0.1-20	79-84		Bachmann et al., 1983 cited in Barber et al., 1999
ABR	8	Synthetic gray water		480	29	0.4	63		Witthauer and Stuckey, 1982 cited in Barber et al., 1999
ABR	6.3	Carbohydrate-protein		8000	35	2.5-36	55-93		Bachmann et al., 1985 cited in Barber et al., 1999
ABR	20	Swine wastewater		5000	30	1.8	75		Yang and Moengangongo, 1987 cited in Barber et al., 1999
ABR	150	Molasses wastewater		5000-10000	37	5.5	98		Yang et al., 1988 cited in Barber et al., 1999
ABR	75	Sucrose		344-500	14	0.7-2	85-93		Orozco, 1997 cited in Barber et al., 1999
ABR	6.3	Distillery wastewater		51,600	30	2.2-3.5	90		Boopathy et al., 1988 cited in Barber et al., 1999

Table 2-1. Anaerobic Bioreactor Studies (continued).

Reactor type	Volume, L	Wastewater Type	Duration, d	Influent COD Concentration mg/L	T, °C	Organic Loading, kg COD m^{-3} d^{-1}	COD Removal Efficiency, %	Comments	Reference
ABR	10	Carbohydrate-protein		4000	35	1.2-4.8	99		Grobicki and Stuckey, 1989 cited in Barber et al., 1999
ABR	9	Carbohydrate-protein		4000	35	1.2-4.8	95		Grobicki and Stuckey, 1991 cited in Barber et al., 1999
ABR	150	Molasses wastewater		116,000	37	4.3-28	49-88		Boopathy and Tilche, 1991 cited in Barber et al., 1999
ABR	150	Molasses wastewater		116,000	37	20	70		Boopathy and Tilche, 1992 cited in Barber et al., 1999
ABR	15	Swine manure		58,500	35	4	62-69		Boopathy and Sievers, 1991 cited in Barber et al., 1999
ABR	350	Municipal waste water		264-906	18-28	2.17	90		Garuti et al., 1992 cited in Barber et al., 1999
ABR	5.16	Slaughterhouse wastewater		450-550	25-30	0.9-4.7	75-90		Polprasert et al., 1992 cited in Barber et al., 1999
ABR	150	Molasses wastewater		116,000-990,000	37	10	40-75		Xing and Tilche, 1992 cited in Barber et al., 1999
ABR	150	Molasses wastewater		116,000-990,000	37	20	70		Xing et al., 1991 cited in Barber et al., 1999
ABR	10	Carbohydrate-protein		4000	35	1.2-4.8	20-80		Nachaiyasit and Stuckey, 1995 cited in Barber et al., 1999
ABR	10	Pharmaceutical wastewater		20,000	35	20	36-68		Fox and Venkatasubbiah, 1996 cited in Barber et al., 1999
ABR	6	Glucose		1000-10000	35	2-20	72-99		Holt et al., 1997 cited in Barber et al., 1999
ABR	10	Carbohydrate-protein		1000-4000	35	1.2-4.8	98		Barber and Stuckey, 1998 cited in Barber et al., 1999
ABR	10	Carbohydrate-protein		4000	15-35	1.2-4.8	75-97		Nachaiyasit and Stuckey, 1997a cited in Barber et al., 1999
ABR	10	Carbohydrate-protein		4000	35	4.8-9.6	90-98		Nachaiyasit and Stuckey, 1997b cited in Barber et al., 1999
ABR	10	Carbohydrate-protein		4000	35	4.8-18	52-98		Nachaiyasit and Stuckey, 1997c cited in Barber et al., 1999
AF	1.5	Municipal supplemented	600	450-780	8	6-20	46-90	Suspended solids removal from 72-95%, lower at low temperature	Bodik et al., 2002a

Table 2-1. Anaerobic Bioreactor Studies (continued).

Reactor type	Volume, L	Wastewater Type	Duration, d	Influent COD Concentration mg/L	T, °C	Organic Loading, kg COD m^{-3} d^{-1}	COD Removal Efficiency, %	Comments	Reference
AF	—	Municipal	—	530	13-15	1.8	35-55		Derycke and Verstraete, 1986 cited in Rebac, 1998
AF	—	Synthetic	—	200	5-10	0.64	27-35		Matsushige et al., 1990 cited in Rebac, 1998
AF	750	Municipal	150	390-440	9-18	0.6-0.75	79-83	Baffled anaerobic filter, SS removal =92%	Bodik et al., 2002b
AF	1.7 x 2	Synthetic	40	500	35	3-8	56-91	Two filters with alternating flow direction	Soedjono and Forster, 1999
AF	4.2	Synthetic	120-180	5000	35	33	90-92		Tay et al., 2000
AF	60	Municipal	140	518	13	3.1	55		Elmitwalli et al., 2002b
AF		Sewage		166-515	25-33	0.8-2.5	71		Raman and Khan, 1978 cited in Ndon et al., 1997b
AF		Sewage		288	20-35	0.288	78		Kobayashi et al., 1983 cited in Ndon et al., 1997b
AMBR	12	Sucrose	78	4000-15500	35	8-30	65-82	Granules	Angenent et al., 2001b
AMBR	20	Milk	185	600	15-20	1.2-3.6	62-85	VFAs increased for high OLR	Angenent et al., 2001a
AMBR	12	Sucrose	120		35	8-23	63-92		Angenent and Dague, 1996
AMBR	20	Sucrose	81		35	10-40	72-85		Angenent et al., 2002
AnSBR	2	Municipal supplemented	720	500-860	8-24	6-20	74-87	Suspended solids removal from 75-92%	Bodik et al., 2002a
AnSBR	—	Synthetic	—	600	5-10	0.6-2.4	65-85		Banik and Dague, 1996 cited in Rebac, 1998
AnSBR	6	Synthetic	142-199	1000	15-35	0.2-1.6	69-98		Ndon et al., 1997b
AnSBR	12	Sucrose	63	3000	35	18.6	80		Angenent et al., 2001b
Hybrid, UASB/AF	60, 65	Municipal	81-144	425-450	13	0.85-1.8	59-71	Higher removal of SS and colloids at 12 hour HRT	Elmitwalli et al., 2002a
Hybrid, UASB/AF	65	Municipal	140	518	13	3.1	34		Elmitwalli et al., 2002b
Hybrid, UASB/AF	4.25	Synthetic		2500-3000	27	5-35	64-96		Guiot and Vandenberg, 1985
UASB	21.5	Screened municipal	180	312 +/- 73	13-25	1.5-2.0	64-72	Methane declined 50%, liquefied COD decreased for low T	Uemura and Harada, 2000

Table 2-1. Anaerobic Bioreactor Studies (continued).

UASB	4.5	Ethanol	12-70	500-10000	25	50	70-85		Jeison and Chamy, 1999
UASB	—	Vinasse	—	200-400	8	0.7-6.5	32-65		De Man et al., 1988b cited in Rebac, 1998
UASB	—	Synthetic	—	1400-7000	10	2-10	49-80		Grant and Lin, 1995 cited in Rebac, 1998
UASB	4.2	Synthetic	120-180	5000	35	11.5	94-97	Shock load study	Tay et al., 2000
UASB	—	Municipal	—	400	8-20	1.20	88-93		Grin et al., 1983 cited in Elmitwalli et al., 2001
UASB		Municipal		467-700	8	1.20-1.25	88-92		De Man et al., 1986 cited in Elmitwalli et al., 2001
UASB		Municipal		465	12-18	0.62	86		Monroy et al., 1988 cited in Elmitwalli et al., 2001
UASB		Municipal		190-1180	12-20	0.57-3.54	84-94		De Man et al., 1988a cited in Elmitwalli et al., 2001
UASB		Municipal		200	18	0.6-1.2	59-63		Sayed and Fergala, 1995 cited in Elmitwalli et al., 2001
UASB		Municipal		507-697	17	4.1-5.6	90		Wang, 1994 cited in Elmitwalli et al., 2001
UASB	12	Sucrose	63	3000	35	19.5	96		Angenent et al., 2001b

Table 2-2. Membrane Coupled Anaerobic Bioreactors Studies.

Scale, Volume, Wastewater Type	Duration, d	HRT/SRT	Organic Loading, kg COD m^{-3} d^{-1} Influent Concentration, mg COD L^{-1} Temperature, °C	Process Type	Membrane Type	Membrane Flux, L m^{-2} h^{-1}	Trans-membrane pressure, kPa	Methane Production, % COD	Effluent Quality	Comments	Reference
Bench-scale, 8.8 L, domestic wastewater with high-strength spikes	90-110	4-6 h/ 150 d (by wast-ing)	3.5-4.2, 100-2600, 12-28	Upflow reactor with sedimentation , packing & suction membrane module	Hollow fiber polyethylene, Mitsubishi Rayon, 0.03 um pore size. Intermittent suction	5-10, runtime = 6d at 10, 14d at 5	5-70, building up over a period of 5 to 14 days	53-66%	92-97% T-COD removal 10-30 mg COD L^{-1}	Effluent quality little affected by loading < 11 kg m^{-3} d^{-1}, or by temperature > 12°C. Significant increase in transmembrane pressure after 4-10 d.	Wen et al., 1999
Pilot-scale, 1600 L each anaerobic & aerobic, raw domestic wastewater	500	4.7-14.2 h/ 72 d	0.44-1.2, 270, 18-27	Mixed anaerobic, Aerated submerged membrane with recirculation	Submerged flat microfiltration membrane modules, polyethylene, 0.4 um pores, Kubota	16-32, required hypo-chlorite & oxalic washing once a year	4-6, max of 17	NA	99% COD, 2.6 mg COD/L	Little buildup of cake on membrane, power consumption = 2.4 kWH per m^3 treated water	Ueda et al., 1999
Pilot plant, 3 m^3, Piggery wastewater, 5000-6000 mg/L, 20 and 35° C. pH 8.4.	85	HRT = 24-48 h acido- & methano-genic	3.75	Separated acidogenic and UASB methano-genic reactors with membrane in 2nd stage	Type NA. 10 min permeation, 5 sec air wash. stainless steel net prefilters, ca. 63 um pore size.	10-38 for 40 d, fell to 0 day 80. 1N NaOH then 1N HCl restored ca 90% flux	NA	0.32 m^3 CH_4/kg COD removed	50-80% COD removal	Prefilter essential to prevent premature caking. Membrane fouled within 10 minutes without backwater. Following consisted of bacteria & mineral deposits.	Lee et al., 2001
Bench-scale, 7 L . Artificial & 2 industrial wastewaters, 5800 -65,000 mg/L COD, pH = 3.2-7.1	75-200	29 h	6-20	Stirred tank reactor with an external filtration device,	Ceramic crossflow membrane, Al_2O_3, 0.2 um pore size.	5-10, crossflow velocity of 2-3 m s^{-1}	NA	0.18-0.34 m^3 CH_4/kg COD removed	80-99% COD removal	High COD removal rates & methane yields, except for slaughterhouse effluent with high particular matter. Membrane allowed retention of slowly biodegradable solids.	Fuchs et al., 2003

Table 2-2. Anaerobic Membrane Bioreactors (continued).

Scale, Volume, Wastewater Type	Dura-tion, d	HRT/SRT	Organic Loading, kg COD m^{-3} d^{-1} Influent Concentration, mg COD L^{-1} Temperature, °C	Process Type	Membrane Type	Membrane Flux, L m^{-2} h^{-1}	Transmem-brane pressure, kPa	Methane Production, % COD	Effluent Quality	Comments	Refer-ence
Bench-scale , 4 L 2-16 L/d unit. 2-16 L/d, 23 g COD/L alcohol distillery wastewater.	199	15	1.5-2.1	Thermo-philic completely mixed anaerobic bioreactor followed by ultrafiltra-tion membrane separation	Plate & frame, DDS Lab., 4 plates in series. DDS Membrane FS61PP, 336 cm^2 surface area, 20000 Dalton cut off. Polypropylene microfilter had higher flux	Initially 70, falling to 20 on day 1, steady decline to 2 by day 140	100-200	NA	500-1500 mg COD/L	SS in bioreactor fell to equal influent solids, Attributed to mixing stress. Organic removal due to activity of biofilm accumulated on the membranes. Major inorganic precipitate was struvite	Choo et al., 1996a, 1996b, 1998
Bioreactors, 5 L, Molasses feed, 20°C,	70-100	NA	0.3-1.0, 1.5-13	Granular bioreactor, Immersed, with suction permeate pump,	polypropylene, 10 um, and fiber glass, 100 um pore size	Fell from 150 to 40 in 40 d	0.1-1.2 suction	NA, Methano-genic activity = ca. 1 g COD g SS^{-1} d^{-1}	Effluent TSS = 10 mg/L for 10 um, 19 mg/L for 100 um	9-24 d cleaning interval, 39-84 d membrane change interval, pH control by bicarbonate addition. Aggregates decreased at 13 kg COD m^{-3} d^{-1}	Hernand ez et al., 2002

NA: not available

Table 2-3. Characteristics of Anaerobic Bioreactor Systems Treating Municipal Wastewaters at Temperatures Less than or Equal to 20 °C that Achieved Effluent CODs Less than or Equal to 65 mg/L.

Type	T, °C	OLR, g/L/d	Infl COD, mg/L	Effl tCOD, mg/L	Effl TSS, mg/L	Vol-ume, L	Reference
AAFEBR	20	0.2-7.7	80-190	30-60	4-20	1	Jewell et al., 1981
AAFEBR	10-20	0.4-1.3	147-196	37-66	2-14	2	Alderman et al., 1998
AAFEBR	10	4.5	300	60	NA	NA	Vanderlast and Lettinga, 1992 cited in Rebac, 1998
AAFEBR	13-15	4.7-9.4	391	16-34	NA	NA	Vanderlast and Lettinga, 1992
ABR	18	2.2	264	26	NA	350	Garuti et al., 1992 cited in Barber and Stuckey, 1999
AF	15	0.8	690	65	14	1.3	Bodik et al., 2002
AF	10	2.1	529	54	NA	NA	Derycke and Verstraete, 1986 cited in Rebac, 1998
UASB	13-16	1.3-1.8	260	50-60	18	21.5	Uemura and Harada, 2000
UASB	8-20	1.2	400	30-50	NA	NA	Grin et al., 1983 cited in Elmitwalli et al., 2001
UASB	8	0.6-1.2	190-700	30-57	NA	NA	De Man et al., 1986 cited in Elmitwalli et al., 2001
UASB	12	4.1	507	51.00	NA	NA	Wang, 1994 cited in Elmitwalli et al., 2002a
Hybrid	13	1.4-1.8	450-533	59-63	64-71	125	Elmitwalli et al., 2002b
MCAB	20	1.6-4.2	400-700	40-65	0	8.9	Wen et al., 1999

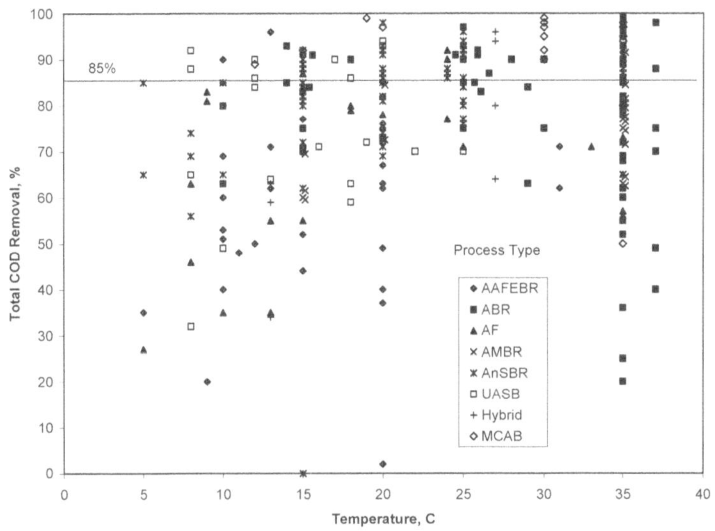

Figure 2-4. Effect of Temperature on COD Removal Efficiency Sorted by Process Type. Data from Studies Listed in Tables 2-1 and 2-2.

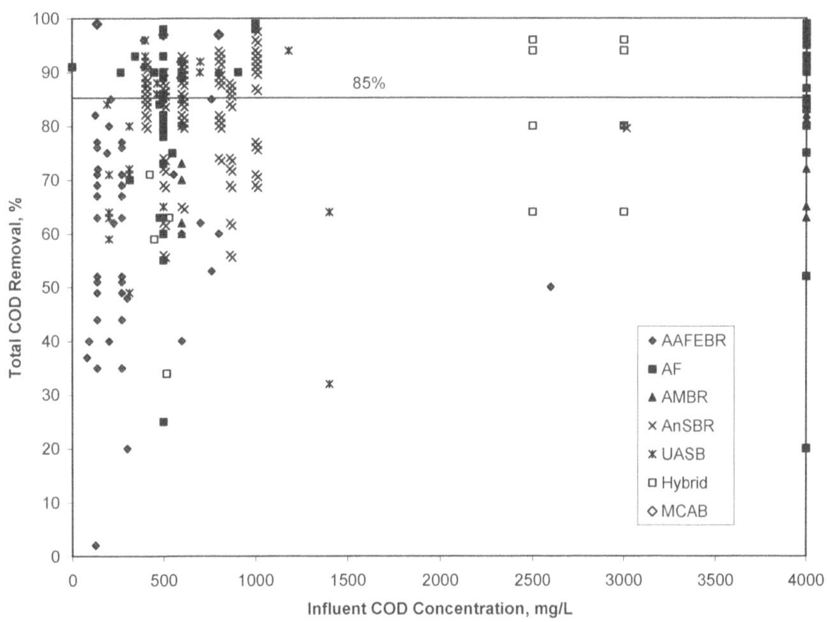

Figure 2-5. Effect of Influent COD Concentration on COD Removal Efficiency Sorted by Process Type. Data from Studies Listed in Tables 2-1 and 2-2.

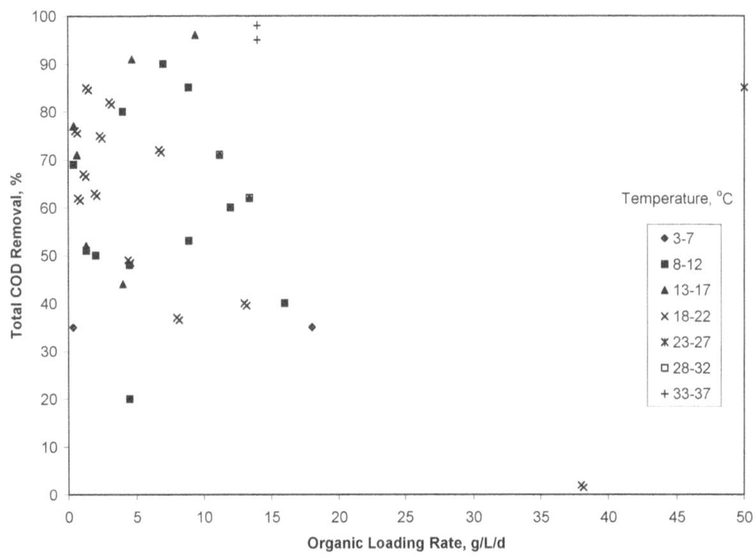

Figure 2-6. Effect of Influent COD Concentration on COD Removal Efficiency for AAFEBR Treatment Studies. Data from Studies Listed in Table 2-1.

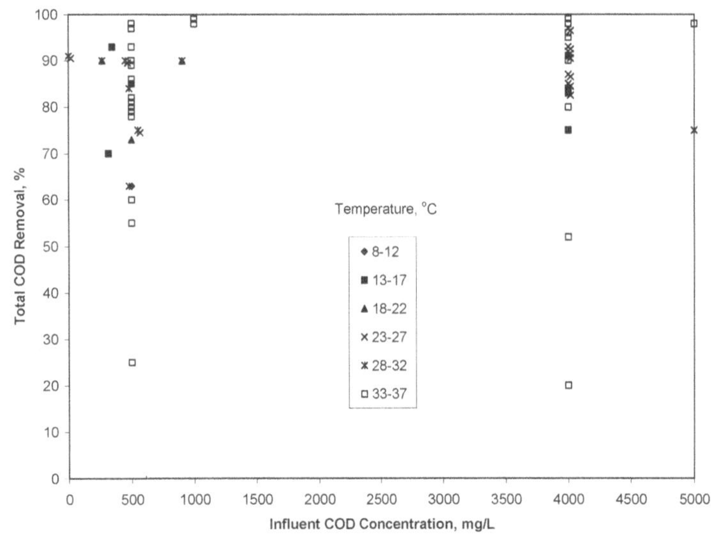

Figure 2-7. Effect of Influent COD Concentration on COD Removal Efficiency for ABR Treatment Studies. Data from Studies Listed in Table 2-1.

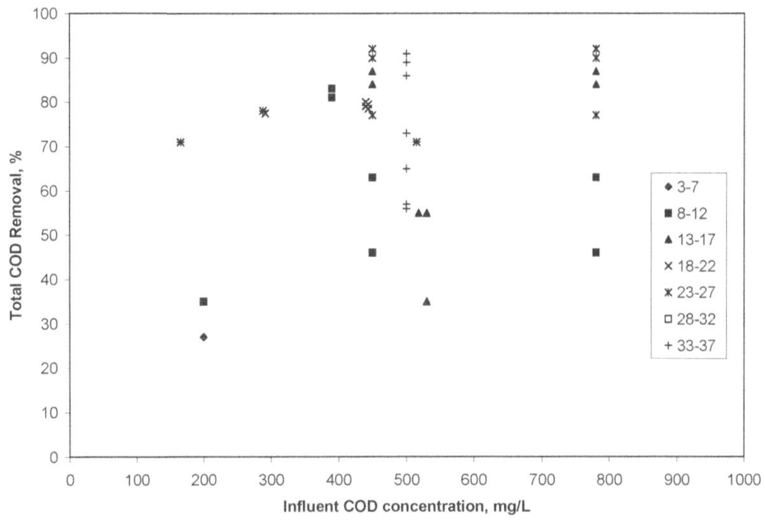

Figure 2.8. Effect of Influent COD Concentration on COD Removal Efficiency for AF Treatment Studies. Data from Studies Listed in Table 2-1.

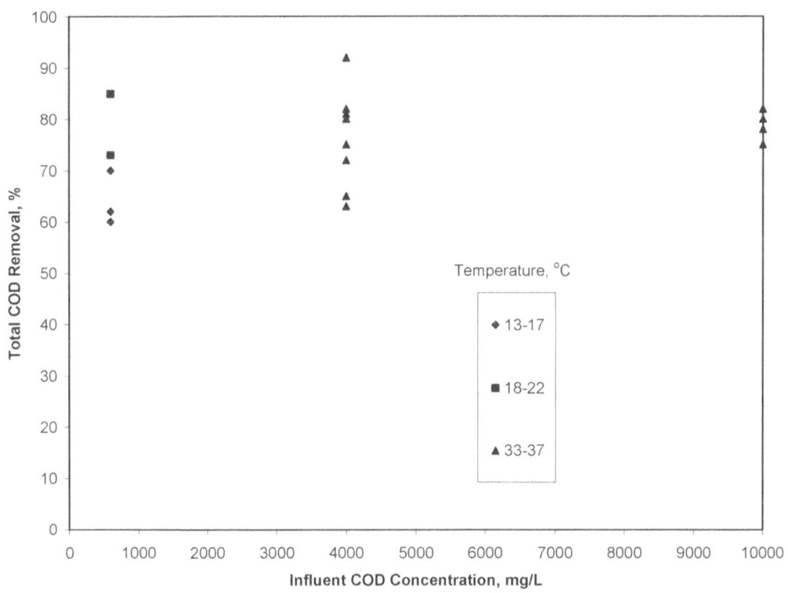

Figure 2-9. Effect of Influent COD Concentration on COD Removal Efficiency for AMBR Treatment Studies. Data from Studies Listed in Table 2-1.

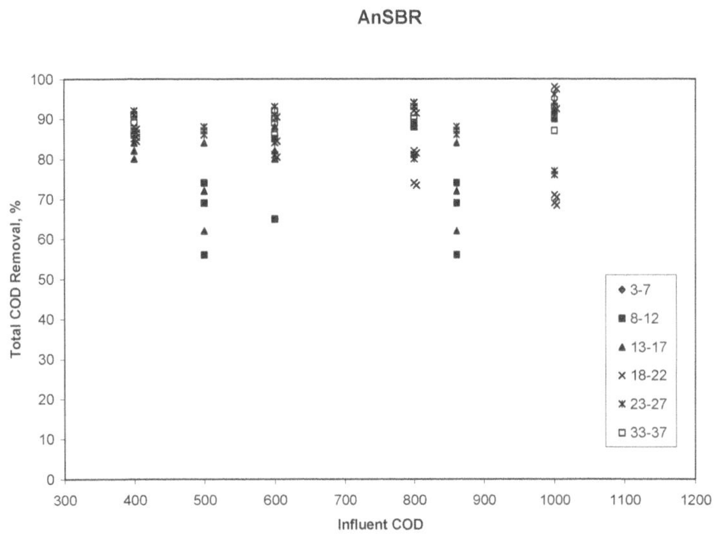

Figure 2-10. Effect of Influent COD Concentration on COD Removal Efficiency for AnSBR Treatment Studies. Data from Studies Listed in Table 2-1.

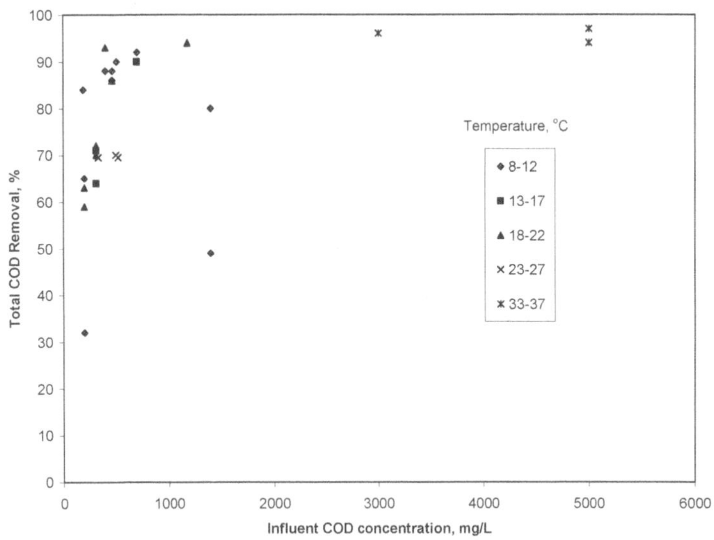

Figure 2-11. Effect of Influent COD Concentration on COD Removal Efficiency for UASB Treatment Studies. Data from Studies Listed in Table 2-1.

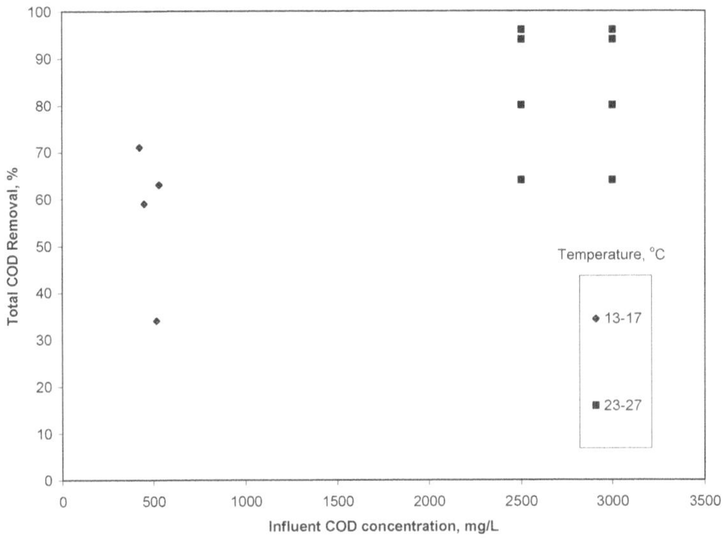

Figure 2-12. Effect of Influent COD Concentration on COD Removal Efficiency for UASB/AF Hybrid Treatment Studies. Data from Studies Listed in Table 2-1.

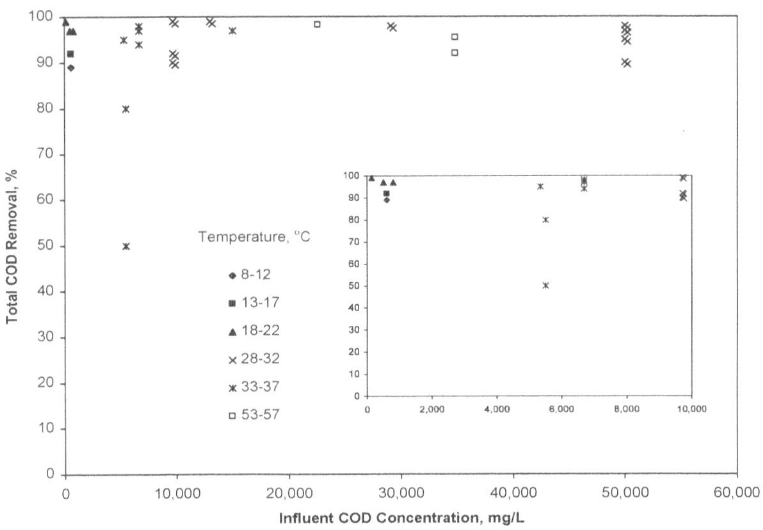

Figure 2-13. Effect of Influent COD Concentration on COD Removal Efficiency for MCAB Treatment Studies. Data from Studies Listed in Table 2-2.

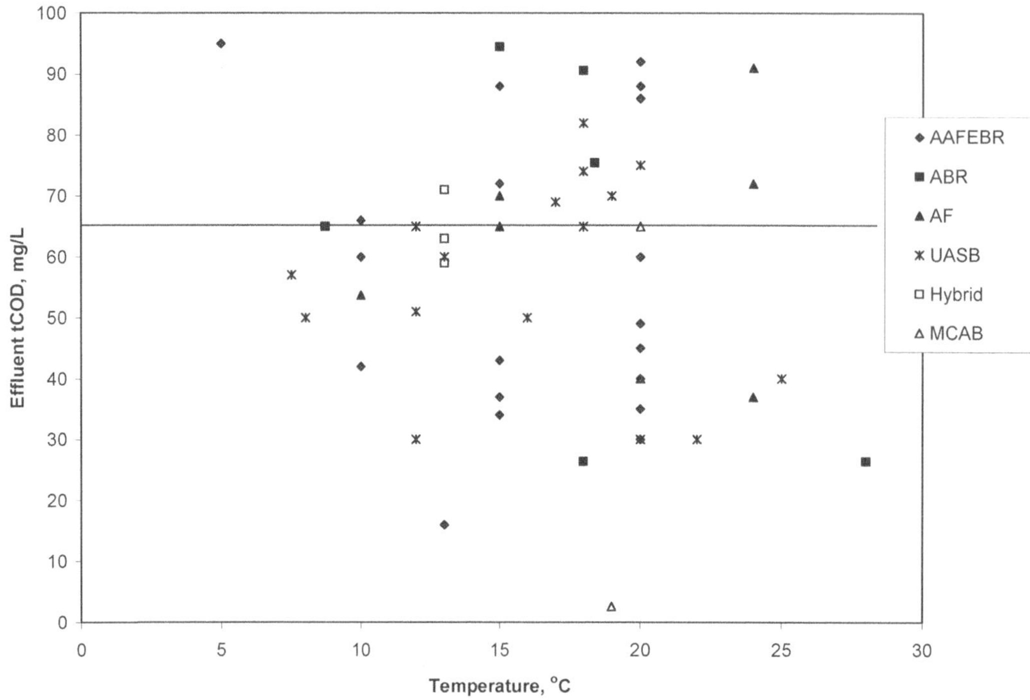

Figure 2-14. Effect of Temperature on Effluent COD Concentration for Municipal Wastewaters Treated Anaerobically. Data from Studies Listed in Table 2-1 and Table 2-2.

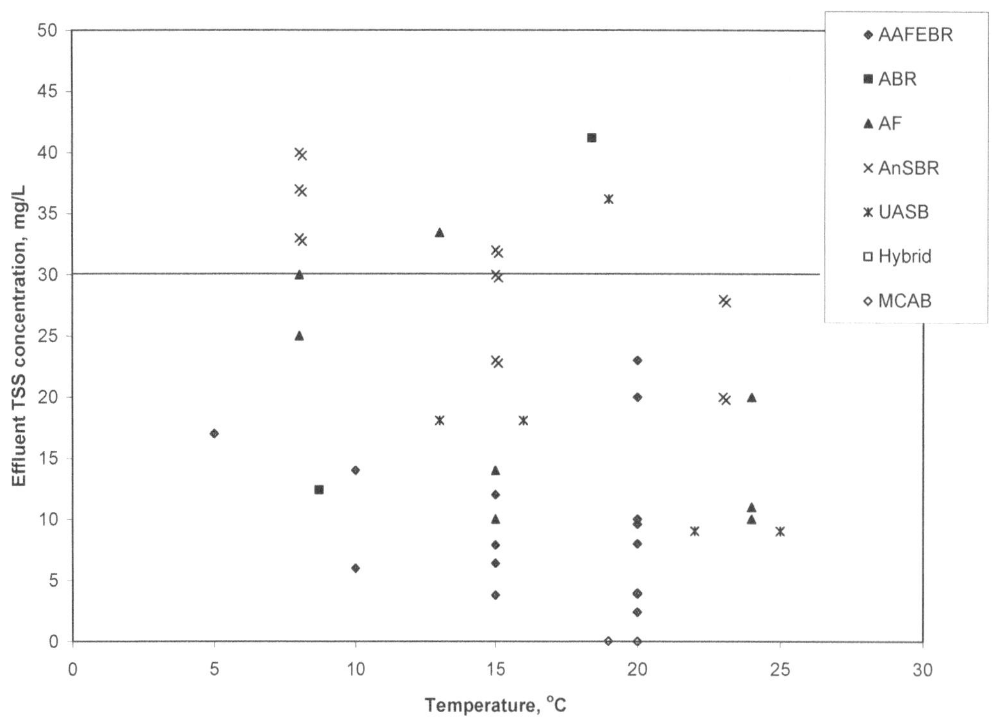

Figure 2-15. Effect of Temperature on Effluent TSS Concentration for Municipal Wastewaters Treated Anaerobically. Data from Studies Listed in Table 2-1 and Table 2-2.

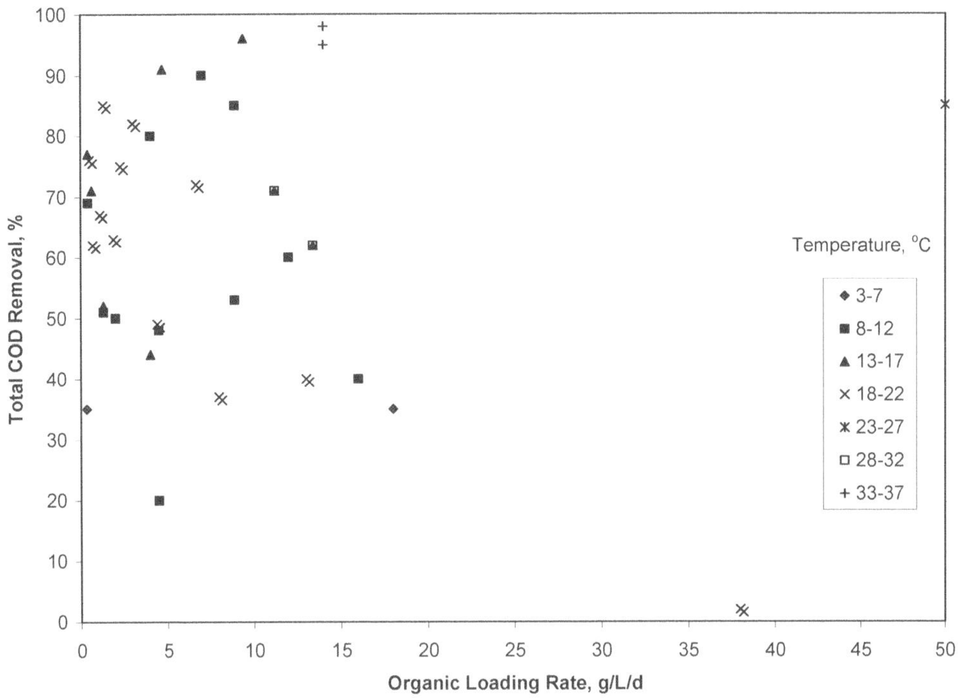

Figure 2-16. Effect of Organic Loading Rate on COD Removal Efficiency for AAFEBR Treatment Studies. Data from Studies Listed in Table 2-1.

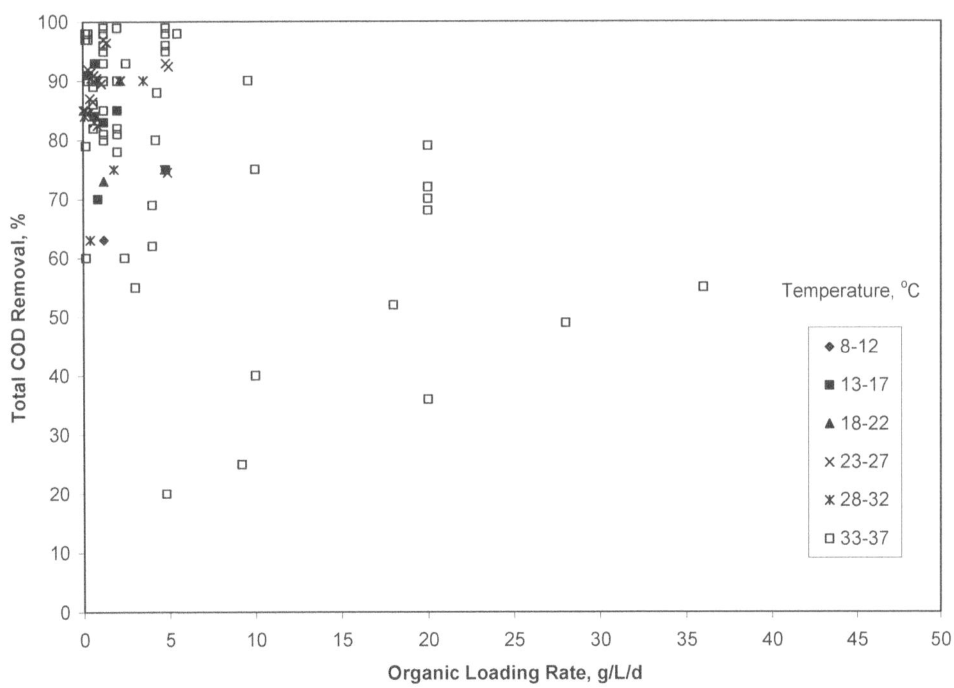

Figure 2-17. Effect of Organic Loading Rate on COD Removal Efficiency for ABR Treatment Studies. Data from Studies Listed in Table 2-1.

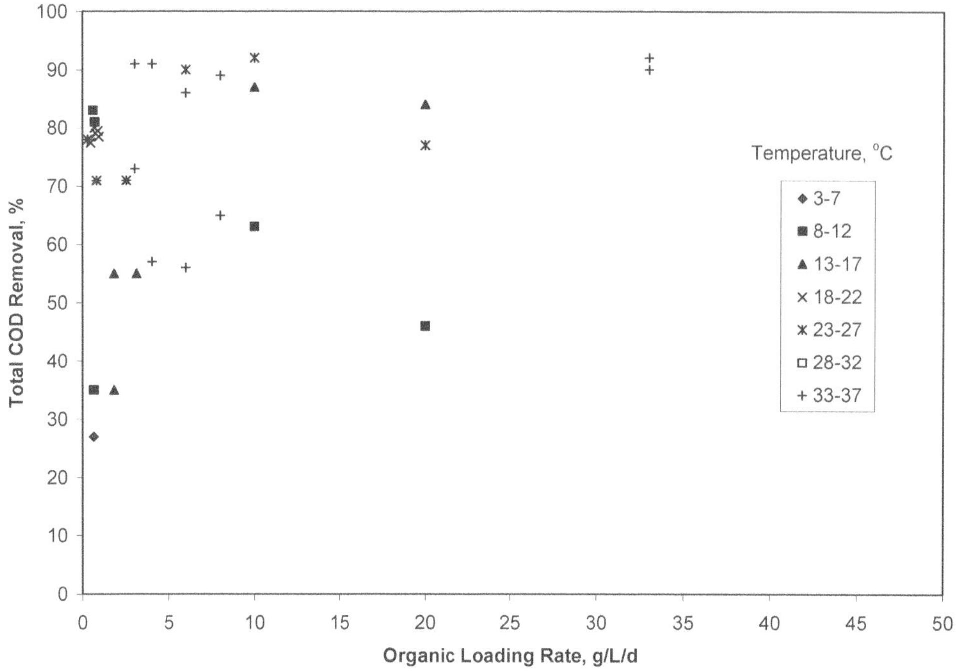

Figure 2-18. Effect of Organic Loading Rate on COD Removal Efficiency for AF Treatment Studies. Data from Studies Listed in Table 2-1.

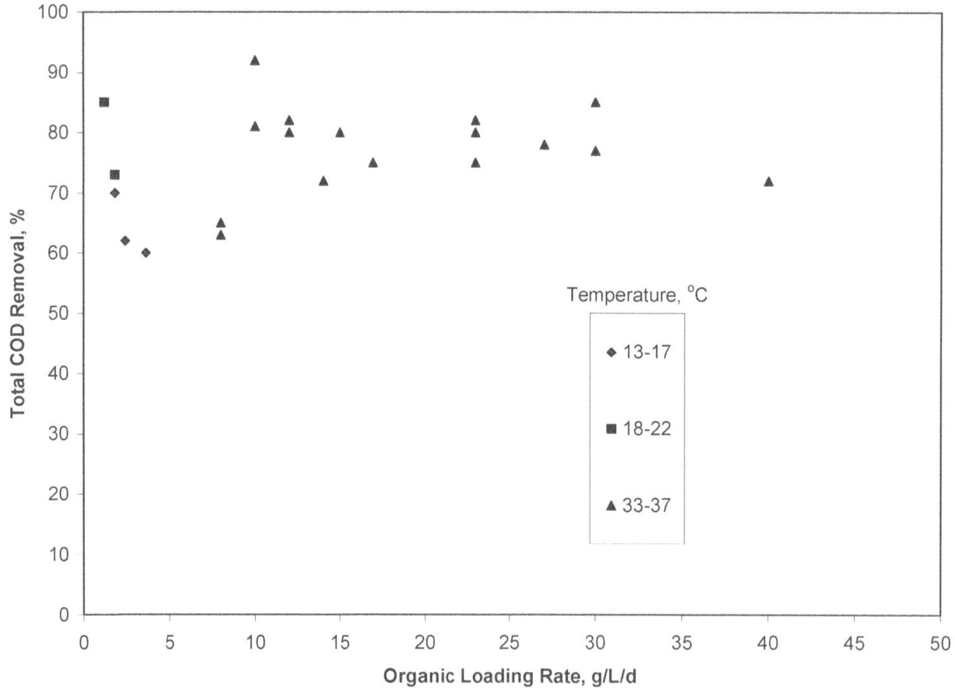

Figure 2-19. Effect of Organic Loading Rate on COD Removal Efficiency for AMBR Treatment Studies. Data from Studies Listed in Table 2-1.

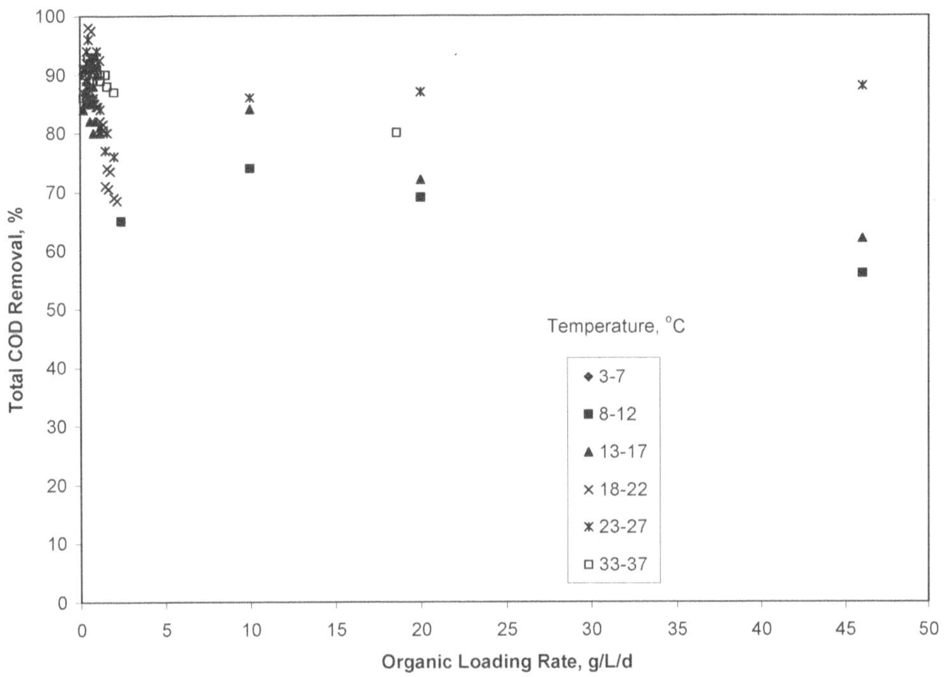

Figure 2-20. Effect of Organic Loading Rate on COD Removal Efficiency for AnSBR Treatment Studies. Data from Studies Listed in Table 2-1.

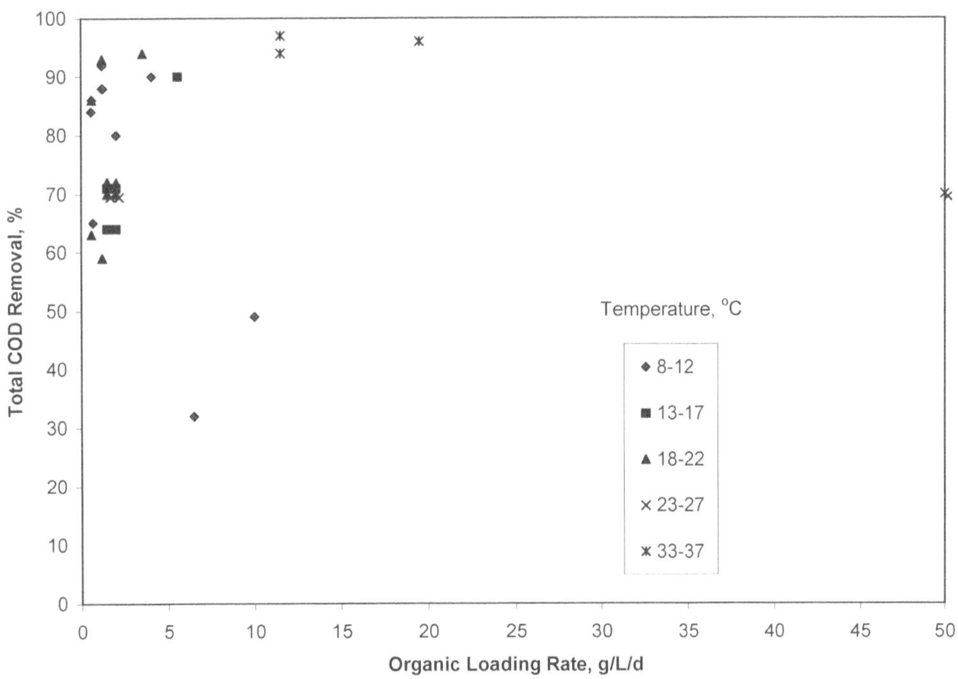

Figure 2-21. Effect of Organic Loading Rate on COD Removal Efficiency for UASB Treatment Studies. Data from Studies Listed in Table 2-1.

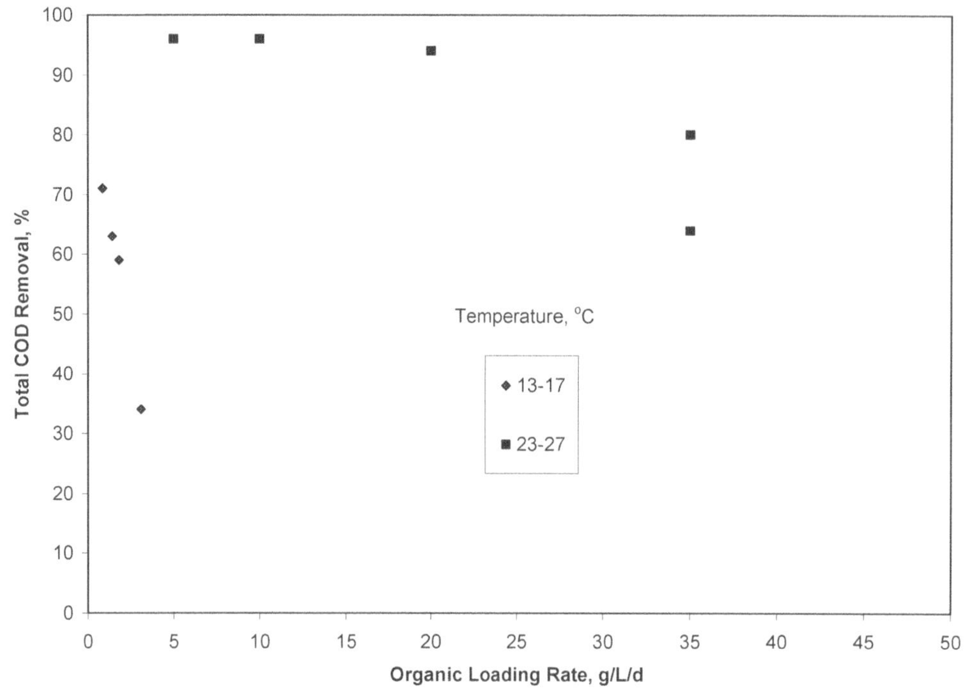

Figure 2-22. Effect of Organic Loading Rate on COD Removal Efficiency for UASB/AF Hybrid Treatment Studies. Data from Studies Listed in Table 2-1.

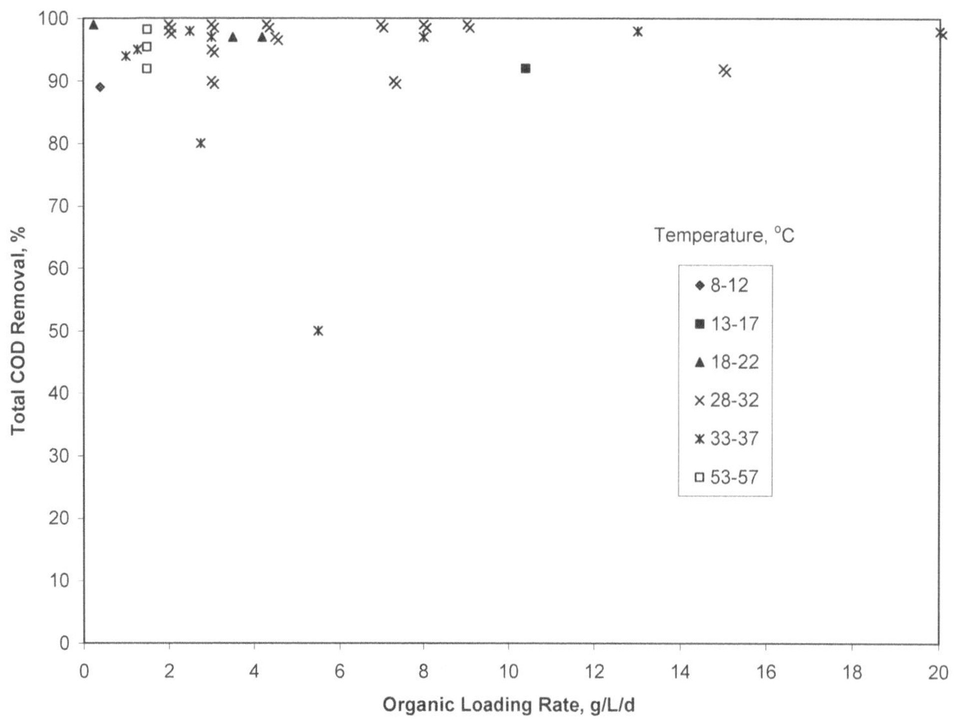

Figure 2-23. Effect of Organic Loading Rate on COD Removal Efficiency for MCAB Treatment Studies. Data from Studies Listed in Table 2-2.

CHAPTER 3.0

SEQUENTIAL ANAEROBIC-AEROBIC TREATMENT

The review of anaerobic processes in Chapter 2.0 showed that most anaerobic treatment applications have been for higher strength industrial wastewaters at warmer temperatures, and in most cases did not need to provide an effluent quality with low BOD and suspended solids concentrations, typical of that needed for secondary treatment of municipal wastewaters. Because anaerobic processes likely would have some residual volatile fatty acids concentration present and tend to produce effluents with more dispersed solids than found for aerobic treatment processes, the ability to produce a secondary effluent quality by anaerobic treatment alone is more challenging than for aerobic treatment processes. The review in Chapter 2.0 supports this expectation, by noting that as the wastewater strength became weaker the COD removal efficiency decreased. Anaerobic processes with membrane separation may improve the effluent quality due to almost 100% solids capture, but alternative sequential anaerobic-aerobic processes may also be considered to meet secondary treatment performance levels. A number of processes have been evaluated at pilot scale to produce a secondary effluent from sequential anaerobic-aerobic treatment processes. These processes are described in Section 3.1.

3.1 Description of Processes

3.1.1 Sequential UASB—Aerobic Treatment Processes

The UASB system has been followed by different aerobic processes to achieve a secondary effluent. The aerobic processes reported in UASB-aerobic systems are: (1) UASB-Biological aerated filters (BAF) (Goncalves et al., 1999; LaCalle et al., 2001); (2) UASB-Sequencing batch reactors (SBR) (Torres and Foresti, 2001); and (3) UASB-Downflow hanging sponge reactors (DHS) (Machdor et al., 2000). A schematic of a UASB-BAF process is shown in Figure 3-1. The BAF process must be designed to remove soluble BOD and suspended solids remaining in the anaerobic treatment process effluent to produce an acceptable secondary effluent discharge. In Figure 3-1 an upflow BAF process is shown, but many variations exist for the BAF process.

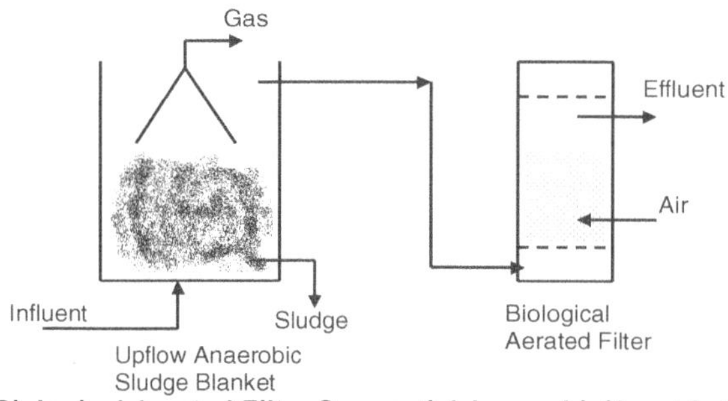

Figure 3-1. UASB-Biological Aerated Filter Sequential Anaerobic/Aerobic Treatment Process.

A BAF is a common term for a fixed film system in which a biofilm is supported in a submerged fixed media with wastewater applied either upflow or downflow. The media size is typically in the range of 4 mm to provide a relatively high volumetric surface area to support biomass growth. Because the media is submerged, oxygen to support microbial growth is supplied by an air sparging system positioned below the media. A filter backwashing system is needed to periodically remove accumulated solids from the filter bed. The backwash water may be collected in holding tanks for sludge thickening and removal. No clarification is used with aerobic submerged attached growth processes, and excess solids from biomass growth and influent suspended solids are trapped in the system and must be periodically removed. Most designs require a backwashing system much like that used in a water filtration plant to flush out accumulated solids, usually on a daily basis.

A number of BAF designs have evolved since their first development in the late 1970s with the Biocarbone process, and these are summarized in Table 3-1 (Water Environment Federation, 2000). The designs differ by their packing material and inlet and outlet flow distribution and collection. The SAFė and Biocarbone processes are similar, in that they both use a downflow operation and 3-5 mm expanded shale media. The heavier media is referred to in the table as a sunken media. The major design changes in the latest generation of BAF processes are the use of a lighter media, with a specific density near that of water (floating media), and the upflow feed application through the packing material. Of the processes shown in Table 3-1, the Biofor, and Biostyr processes are more commonly used. The main differences between the Biofor and Biostyr processes are their media composition and density. The media size used (3–5 mm) and media depths (3–4 m) are generally similar. The Biostyr process uses a polystyrene bead media that is slightly less dense than water. Because of the upflow direction, a plate is located above the media to retain it and contains nozzles for collection of treated effluent. The Biofor process uses the heavier fired clay material, which remains at the bottom of the BAF reactor and is referred to as a "sunken" media. To protect the nozzles that distribute wastewater upflow through the Biofor media, fine screening of the influent is necessary. Both processes use similar backwashing procedures of air and water.

Table 3-1. Summary of Biological Aerated Filter Systems.

Process	Manufacturer	Flow Direction	Support Medium	Type of Medium	Process Air Direction
Biobead	Brightwater (with Anglia Water Company)	Upflow	Polyethylene	Floating	Cocurrent
Biocarbone	OTV	Downflow	Expanded Shale	Sunken	Counter-Current
Biofor	Ondeo Degremont	Upflow	Biolite	Sunken	Cocurrent
Biostyr	U.S. Filter	Upflow	Polystyrene	Floating	Co-current
ColOX	Tetra	Upflow	Sand	Sunken	Counter-Current
SAFe	PTW Projects (Thames Water Company)	Downflow	Expanded Shale	Sunken	Counter-Current
Stereau	Purac	Downflow	Pumice	Sunken	Counter-Current

3.1.2 UASB-Sequencing Batch Reactor (SBR) Process

A SBR system provides an alternative form of activated sludge treatment after a UASB reactor. For a SBR, a single tank is used for both aeration and effluent clarification. The reactor is operated in sequential time periods of fill, aeration, settling, and effluent decanting. An anoxic or anaerobic mix step may also occur during the fill period and for a period prior to aeration to effect nutrient removal. Process advantages are operational flexibility, economical for small facilities, potential for nutrient removal, quiescent conditions during solids-liquid separation, and ability to handle wide variations in flow and loadings (Tchobanoglous et al., 2003).

3.1.3 UASB-Downflow Hanging Sponge (DHS) Process

The downflow hanging sponge (DHS) reactor after a UASB reactor represents a unique type of unsubmerged fixed-film process investigated at Naguoka University of Technology in Naguoka, Japan (Machdor et al., 2000). The downflow reactor consists of a series of connected hanging sponge-cubes in a 0.8–1.6 cm depth. The cubes consist of 3 cm by 75 cm polyurethane foam. Oxygen is supplied by diffusion from the airspace between the sponge and vertical sponge curtains. No liquid-solids clarification unit was used after the aerobic fixed film reactor.

3.1.4 Anaerobic/Aerobic Treatment with Nitrogen Removal

The ANANOX (Anaerobic-Anoxic-Oxic) process, shown in Figure 3-2, consists of an anaerobic baffled reactor (ABR) for the anaerobic treatment phase, followed by an aerobic activated sludge process. The process is intended to accomplish nitrogen removal in addition to secondary treatment. A portion of the nitrified effluent is recycled to the last stage of the three-staged ABR, so that nitrate can be biologically reduced in the anaerobic environment. An approach to combine nitrogen removal with an anaerobic-aerobic process was also evaluated by LaCalle et al. (2001) for a UASB-BAF system. Nitrified effluent from the BAF process was recycled to the influent of the UASB reactor to accomplish biological denitrification.

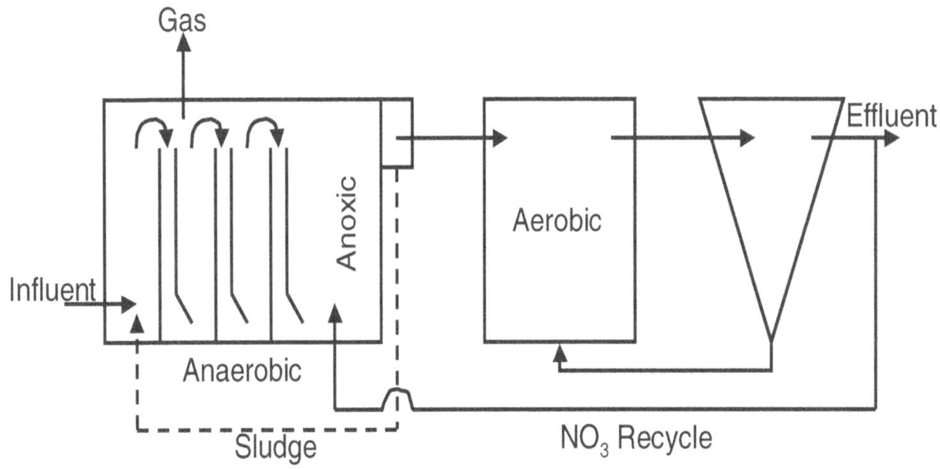

Figure 3-2. The ANANOX Sequential Anaerobic/Aerobic Treatment Process for COD and Nutrient Removal.

3.2 Process Fundamentals and Biomass Reduction Mechanisms

The ability to significantly reduce biomass production depends on the degree of anaerobic treatment in the first phase of the sequential anaerobic-aerobic processes, and is independent of the aerobic step. Thus, conditions described in Chapter 2.0 for anaerobic treatment—such as temperature, wastewater strength and characteristics—and process design affect the anaerobic reactor efficiency and biomass reduction performance. The more organic conversion that occurs in the anaerobic step, the greater the system biomass reduction will be. The aerobic treatment step is intended as a secondary treatment polishing step to remove BOD and suspended solids to acceptable levels.

3.3 Reported Process Performance

Results from specific studies with the sequential anaerobic-aerobic processes are summarized in this section for each of the processes described.

A UASB-BAF process was studied with bench-scale reactor operations at >20 °C for over 320 days, treating raw domestic wastewater from Vitoria, Brazil (Goncalves et al., 1999). The UASB reactor was 10 cm diameter with a 4.6 m height, and the upflow BAF unit was 10 cm diameter with a 0.8 m height containing 3.0 mm polystyrene beads. Testing was done at successive lower hydraulic retention times (HRTs), and average results are summarized in Table 3-2. At HRTs of 6 to 8 hours, the UASB process had an effluent BOD and TSS concentration that was only a few mg/L higher than that for a typical secondary effluent of 30 mg/L BOD and TSS concentration. In all cases the BAF process polished the effluent to BOD and TSS concentrations at about 10 mg/L.

The solids production was reported in the range of 0.15 kg TSS/kg COD removed (COD_r) for the anaerobic step and 0.4 kg TSS/kg COD_r for the BAF step. Thus, a significant reduction in biomass production compared to conventional systems was demonstrated. Considering that 80–85% of the influent BOD was removed in the anaerobic step, the net solids production for liquid processing was about 0.20 kg TSS/kg COD, which is less than half of the expected solids production for conventional wastewater treatment (Table 1-2).

A bench-scale UASB-BAF system was evaluated for a food industry wastewater with average influent COD, BOD, TSS, and TKN concentrations of 10,400; 5,250; 1,275; and 790 mg/L, respectively (Lacalle et al., 2001). The UASB and BAF reactor volumes were 10 and 3 L, respectively, and the system was operated at 33 °C. The BAF unit contained sunken media consisting of 4 mm fired clay material. Nitrified effluent from the BAF reactor was recirculated to the UASB influent at recycle flow to influent flow rate ratios from 6.7 to 10.4.

Table 3-2 Average Treatment Performance for a UASB-BAF Process Treating Raw Domestic Wastewater in Bench-Scale Study at > 20 °C in Vitoria, Brazil (Goncalves et al., 1999).

PHASE	Reactor	BOD loading kg BOD/m^3-d	HRT (hrs)	Average Effluent, mg/L			
A	UASB BAF	0.7 1.2	8.0 0.7	112 49	36 10	35 26	37 10
B	UASB BAF	0.9 1.4	6.0 0.5	105 38	29 9	33 27	35 10
C	UASB BAF	0.8 1.7	4.0 0.3	88 49	21 10	33 28	30 10

More than 90% nitrogen removal was achieved, and final effluent nitrogen concentrations averaged 70 mg N/L (TKN and NO_x-N at 13 and 57 mg/L, respectively). The system loadings averaged 2.0 kg COD/m^3-d. The UASB effluent TSS concentration was about 170 mg/L, but the BAF effluent TSS concentration was reported to be maintained near 10 mg/L. The average solids yield reported for the system was 0.1 kg VSS/kg COD_r. Thus, this study demonstrated the ability to achieve a significant reduction in solids production, nitrification and denitrification with the sequential system, and the use of an aerobic polishing step after the anaerobic process.

The UASB-SBR system was also evaluated at higher temperatures (21°C) in a pilot system treating raw domestic wastewater at Sao Carlos City, Brazil (Torres and Foresti, 2001). The UASB unit was operated with a 6-hr HRT and the activated sludge SBR process was operated with a 30-day SRT and had equivalent HRTs (volume divided by daily flow treated) of 0.5 to 3.0 days as the feed cycles/day varied in different tests from 6 to 1, respectively. Average treatment performance is summarized in Table 3-3. The SBR process was able to provide nitrification at the lower number of cycles per day and subsequently lower loadings. A secondary effluent quality was normally achieved treating the UASB effluent. Details were not provided on the SBR mixed liquor concentrations or amount of nitrate removal during the fill period. The authors noted low solids production, but did not provide specific net solids yield values. Based on the relative COD removals in the UASB and SBR units, the net biomass production was theoretically 50–60% of that for aerobic treatment only.

The UASB-DHS system operated by Machder et al. (2000) also was maintained at elevated temperatures (25 °C). A 155-liter UASB unit treated domestic wastewater at an average HRT of 6 hours and the DHS unit had an HRT, based on empty bed volume, of 2.0 hrs. The average amount of COD and BOD removed in the UASB reactor was 58 and 66%, respectively.

Based on these removal fractions, the biomass production was theoretically about 55–60% of that for a conventional system. Though the treatment results for the DHS unit showed high soluble BOD conversion (88–94%), the effluent suspended solids were high due to the lack of an efficient liquid-solids separation unit following biological treatment.

Table 3-3. Average Treatment Performance for UASB-SBR Pilot Plant Treating Raw Domestic Wastewater at 21 °C at Sao Carlos City, Brazil (Torres and Foresti, 2001).

	Concentration, mg/L			
Parameter	**Influent**	**UASB**	**1 cycle/d**	**4 cycle/d**
COD	569	167	13	15
TSS	131	49	6	8
TKN	42	40	5	8
NH_4-N	23	39	< 1	< 1
PO_4	14	12	11	7

The ANANOX process (Figure 3-2) was demonstrated in a pilot plant operation at 22 °C with a total reactor volume of 700 L treating screened domestic wastewater at the Bologna, Italy municipal wastewater plant (Garuti et al., 1992). The HRTs of the two- staged baffled upflow anaerobic reactor, the subsequent anoxic upflow reactor, and the aerobic reactor were 4.7, 3.7, and 8.6 hours, respectively. The organic loading to the anaerobic stage was 2.2 kg COD/m^3-d. The average performance over a six-month period is summarized in Table 3-4. Only about 40% COD removal occurred in the anaerobic zone. Based on the influent NH_4-N concentration, 77% nitrogen removal occurred, but the efficiency is higher if the influent organic nitrogen (measured by TKN) is included. Overall the system achieved a secondary effluent quality with nitrogen removal by nitrification and denitrification. The observed solids production was estimated at 0.2 g TSS/g COD_r which is 40–50% of that expected for liquid treatment by biological aerobic-only processes.

Table 3-4. Average Treatment Performance for an ANANOX Process Pilot Plant Treating Screened Domestic Wastewater at 22 °C for a 6-Month Period (Garuti et al., 1992).

Parameter	**Influent, mg/L**	**Effluent, mg/L**		
		Anaerobic	**Anoxic**	**Aerobic**
TCOD	428	258	196	43
SCOD	157	91	45	21
TSS	250	169	196	27
NH_4-N	47	34	17	1.3
NO_3-N	1.6	1.0	2.1	10.4

3.4 Plant Retrofit Considerations

The review of systems used in anaerobic-aerobic processes showed that different types of aerobic systems could be used after the anaerobic process, including activated sludge under different configurations, and fixed film processes (BAFs). Thus, it is reasonable to expect that most existing aerobic biological treatment systems would be amenable to the addition of an anaerobic pretreatment step. The tank requirements, piping, and layout consideration would be site-specific. For cases where the anaerobic process is more efficient with relatively low suspended solids and BOD concentrations, fixed film processes should be considered as it may be difficult to maintain a sufficient mixed liquor concentration in an activated sludge process. Though not tested, a trickling filter operation also should be feasible after an anaerobic process. The DHS process was closer to that type of fixed film treatment and showed that a liquid-solids separation step (e.g., a secondary clarifier with chemical addition) still would be needed.

3.5 Assessment of Process

The assessment of the sequential anaerobic-aerobic process is dependent on the performance of the anaerobic process (Chapter 2.0). The anaerobic process effluent can be treated to a secondary process effluent by most existing aerobic biological treatment processes. The key issue for the process feasibility, as indicated in Chapter 2.0, is the ability of the anaerobic process to perform effectively for lower strength wastewaters as a function of temperature.

The studies presented in Section 3.3 indicated that high COD conversion can be achieved in the anaerobic process for domestic wastewater at temperatures above 20 °C. Thus, these processes may be very applicable for domestic wastewater treatment at warmer climate locations in southern parts of the United States mainland and for Hawaii.

A site-by-site design and economic analysis would be needed to assess the competitiveness of the process versus conventional treatment. The total tank volume would conceivably be greater for a coupled anaerobic-aerobic process, but other costs would be less. The sludge production could be less by 50% or more resulting in lower costs for sludge processing, hauling, and disposal/reuse. Energy would be produced in the form of methane, and less energy would be consumed for aeration than with conventional aerobic biological processes. For most designs no additional sludge digestion units would be needed. As an example, Table 3-5 (Goncalves et al., 1999) compares a conventional aerobic wastewater treatment process to an anaerobic-aerobic UASB-BAF process for a system serving a population of 25,000 in a warm climate. The comparison suggests that an anaerobic-aerobic process can be competitive with conventional systems for warm climates.

The reported test results in Section 3.3 showed that nitrogen removal is feasible with anaerobic-aerobic processes. The aerobic process can be designed and operated to achieve biological nitrification. A recycle stream from the aerobic process must be used to contact nitrate with influent BOD or anaerobic effluent BOD. Further demonstration and evaluation of this type of process is needed to assess its reliability, as a certain potential problems are apparent. First, the recycle can increase the hydraulic load to the anaerobic process, which could affect its treatment efficiency. For the alternative mode of feeding nitrified recycle after the anaerobic step, variations in the anaerobic step performance could possibly result in insufficient BOD for denitrification of the recycled nitrate. For some designs (as shown for the ANANOX process) where it is necessary to recycle after the final clarification step, the recycle will also increase the hydraulic loading to be handed by the biological treatment units, such as clarifiers.

Table 3-5. Comparison of a Conventional Aerobic Process to an Anaerobic/Aerobic Process for Domestic Wastewater Treatment (25,000 People).

Conventional		Anaerobic-Aerobic	
Processes	**Volume, m^3**	**Processes**	**Volume, m^3**
Primary clarifiers	300	UASB	1200
BAFs	233	BF	90
Thickener	96		
Digesters	550		
Total	1180		1290
Operation Costs	310	237	% of conventional
Sludge production, kg TS/d	49	14	76
Power, kW	174	320	29
Methane produced, kg/d			184

3.6 Research Needs

The key research need for the anaerobic-aerobic process is with the anaerobic process. Research is needed to develop designs that can provide reliable performance at low temperature for weak to moderate strength wastewaters. These conditions may require the use of membrane separation with anaerobic processes to maximize the methanogenic biomass concentration and SRT in order to increase the efficiency of converting organics to methane gas and minimize biomass production.

An additional research need is in the area of nutrient removal. Further work should be done to evaluate nitrification-denitrification processes that integrate the anaerobic and aerobic processes in sequential anaerobic-aerobic systems. Such work should be done at the pilot scale and should address long-term performance and control issues associated with variations in performance for the anaerobic treatment step.

CHAPTER 4.0

ANAEROBIC-AEROBIC ENHANCED BIOLOGICAL PHOSPHORUS REMOVAL PROCESSES

4.1 Process Description

The advent of a new generation of activated sludge treatment technology occurred in the mid 1970s when Barnard (1974) recognized key process requirements that promoted enhanced biological phosphorus removal (EBPR) in activated sludge processes. The key process configuration, termed the Phoredox process, involved the use of an anaerobic contact basin with a relatively short hydraulic retention time (45–60 minutes) prior to the aerobic treatment reactor (Figure 4-1). This process is commonly referred to as the A/O process today.

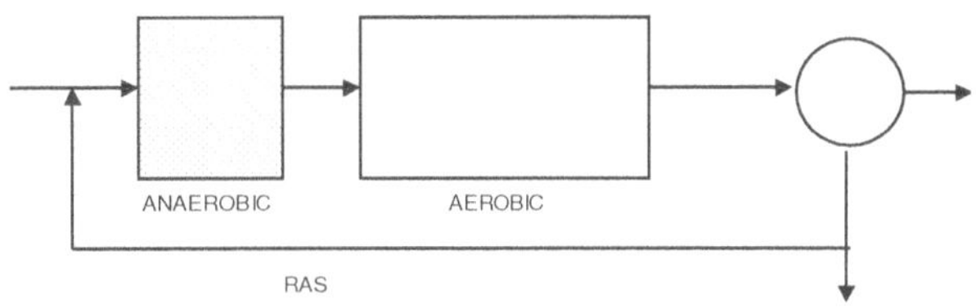

Figure 4-1. Anaerobic/Aerobic Process (A/O) for Enhanced Biological Phosphorous Removal (EBPR).

Influent wastewater and return activated sludge (RAS) are mixed in the anaerobic reactor prior to an aeration tank. Nitrate and oxygen input to the anaerobic reactor must be minimal, so that one of the main metabolic conditions for substrate utilization is anaerobic fermentation. Barnard (1974) noted that nitrate or oxygen in the anaerobic contact zone would consume substrate that would otherwise be used for EBPR. The process configuration promotes the selection of bacteria that are capable of removing and storing phosphorus at high levels in the aeration zone, and are referred to as phosphorus accumulating organisms (PAOs). Phosphorus is removed from the system via excess sludge production rich in stored phosphorus. Other process configurations have been developed to maximize phosphorus removal efficiency using the concept of anaerobic contacting preceding aerobic treatment.

Other process configurations that incorporate the A/O process are presented in Figures 4-2 through 4-5. They show different designs that were developed to minimize the amount of nitrate that may enter the anaerobic contact zone in the RAS. For systems in which nitrification occurs, the simple A/O treatment scheme shown in Figure 4-1 is not appropriate because the nitrate in the RAS flow will result in decreased EBPR efficiency. For the anaerobic/anoxic/aerobic (A^2O) process in Figure 4-2, 75–85% of the nitrate produced by nitrification in the aerobic zone is removed by an internal recycle of nitrate to a preanoxic zone, where nitrate provides an electron acceptor for BOD removal in lieu of aerobic respiration.

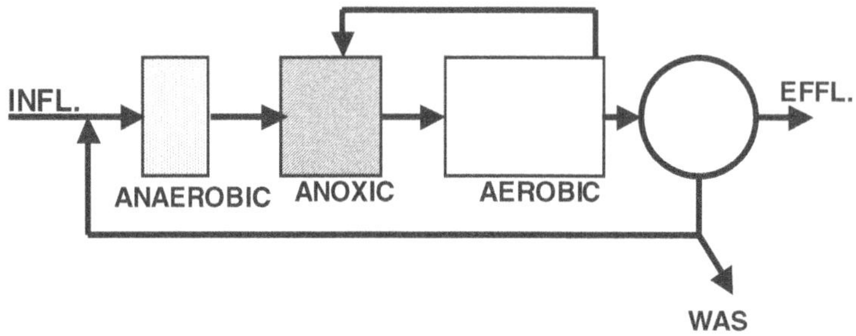

Figure 4-2. Anaerobic/Anoxic/Aerobic A^2O Process for EBPR with Nitrification.

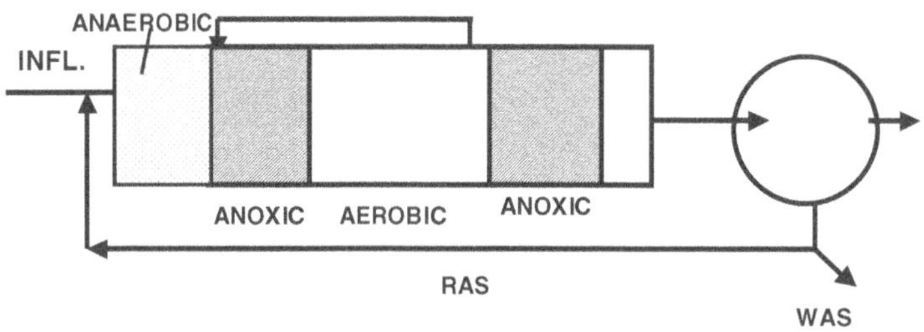

Figure 4-3. Modified Bardenpho Process for EBPR and to Minimize Nitrate.

Thus, a lower concentration of nitrate is directed to the anaerobic reactor in the RAS stream. In the modified Bardenpho Process (Figure 4-3), the RAS nitrate concentration is minimized further by the addition of an anoxic zone after the aeration zone.

Processes that result in de minimus nitrate feed to the anaerobic zone are described by Figures 4-4 and 4-5; the University of Capetown (UCT) and Johannesburg (JHB) processes, respectively. In the UCT process the RAS is directed to an anoxic zone first for nitrate removal. The anoxic zone mixed liquor, with a minimal nitrate concentration, is recycled to the anaerobic zone to provide mixed liquor to the reactor. Another version of the UCT process is the modified UCT (MUCT) process in which the anoxic zone is staged with return of mixed liquor from the last stage to the anaerobic contact zone to assure minimal nitrate to the anaerobic zone.

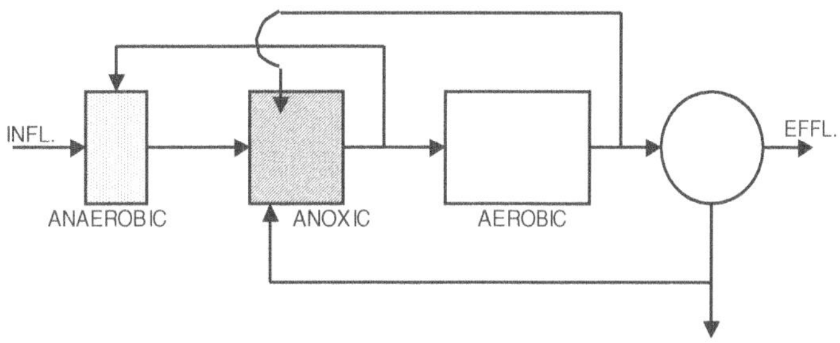

Figure 4-4. University of Capetown (UCT) Process for EBPR with Minimal Nitrate Feed to the Anaerobic Zone.

The JHB process (Figure 4-5) has a simpler operation and less recycle systems than the UCT or MUCT, but also assures minimal nitrate feed to the anaerobic contact zone by holding the RAS in a mixed anoxic tank prior to the anaerobic zone. All of these processes have the necessary anaerobic contact zone to provide EBPR.

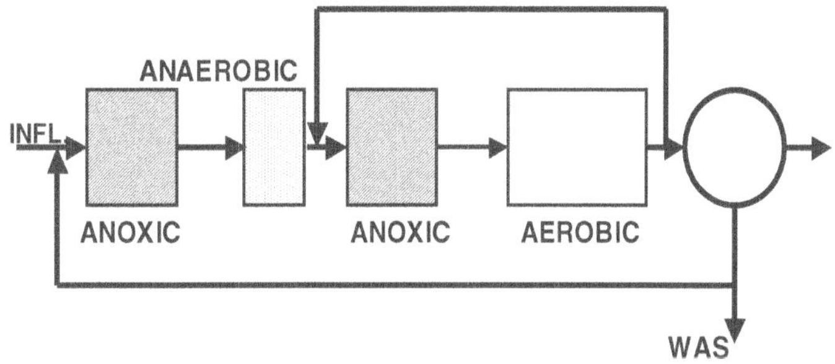

Figure 4-5. Johannesburg Process for EBPR to Minimize Nitrate in Influent to the Anaerobic Contact Zone by Denitrification of RAS.

Understanding the process steps in the anaerobic and anoxic/aerobic zones of an EBPR system provides the necessary background for the proposed COD loss mechanism that could lead to less biomass production than with a conventional aerobic-only activated sludge process. Important biological reactions for the anaerobic contact zone are summarized in the following (Tchobanoglous et al., 2003). Acetate is a key substrate; some influent wastewaters contain acetate, but the majority of the acetate usually is produced in the process. Acetate is produced by fermentation of soluble organic compounds that come from the influent wastewater, referred to as readily biodegradable COD (rbCOD), and from hydrolysis of biodegradable particulate and colloidal COD. Because of the relatively short detention used in the anaerobic contact zone, most of the acetate is derived from influent rbCOD. Using energy available from stored polyphosphates, the PAOs assimilate the acetate and produce intracellular storage carbon products such as polyhydroxybutyrate (PHB). Concurrent with acetate uptake is the release of orthophosphorus, as well as magnesium, potassium, and calcium ions. The intracellular PHB content in the PAOs increases, while the polyphosphate concentration decreases. Glycogen also is consumed within the PAOs cells during PHB formation.

In the subsequent anoxic and aerobic zones the following occurs. The stored intracellular storage products are oxidized by the PAOs to provide energy and a portion is also used for new cell growth. Some glycogen is produced from PHB degradation and is available later in the anaerobic zone. The energy produced from PHB oxidation is captured as the PAOs take up soluble orthophosphate to produce polyphosphate storage products. As a portion of the biomass is removed in the waste sludge for the system, phosphorus removal occurs via wasted solids.

In early studies on the A/O EBPR process, Bordacs and Tracy (1988) observed that the oxygen requirements in the aerobic zone following the anaerobic contact zone were about 30% less than expected for a conventional activated sludge process. Their explanation was that stored intracellular carbon products were degraded more slowly in the EBPR process than for alternative BOD oxidation pathways in conventional processes. This explanation would suggest a higher biomass yield for EBPR systems at a given SRT due to a less complete oxidation of removed COD. Later, Randall et al. (1992) reported an unexplained COD less in EBPR A/O systems, and claimed that the COD loss was occurring in the anaerobic contact zone. A few other investigators subsequently made similar claims.

The occurrence of a COD loss from EBPR systems suggests both reduced oxygen needs and less sludge production. Conventional thinking on activated sludge process design is that when reduced oxygen demand occurs, biomass production increases and vice versa. Because of the unique nature of COD loss claimed for EBPR systems, the lack of direct proof of the COD loss mechanism, and the lack of any conclusive side-by-side studies with a control and designs that are thought to promote COD loss, further study is warranted within this project scope to investigate the feasibility of biomass reduction processes.

4.2 Process Fundamentals and Biomass Reduction Mechanism(s)

4.2.1 Reported COD Loss

A comprehensive summary of previous experiments, and the results of an experiment using an A/O and aerobic-only process in parallel control reactors have been summarized by Randall et al. (1992) to show COD loss for A/O processes. In all of these studies the COD loss claims were based on a comprehensive COD mass balance for laboratory or pilot-scale reactors that required numerous measurements. The measurements included the total daily amount of COD removed across the system based on COD in and out (including effluent liquid COD and wasted solids). The COD removal across the system should be equal to the oxygen consumption for oxidation of carbonaceous COD, which was determined from oxygen uptake rate test data and nitrogen balances. The carbonaceous oxygen demand was found by correcting the measured biological oxygen uptake rate data by subtracting the amount of oxygen used for nitrification and adding the oxygen equivalent supplied by the reduction of nitrate and nitrite in the anaerobic and anoxic zones. An important component of the COD balance employed was the oxygen equivalent of the waste solids, which was estimated at 1.4 g 0_2/g VSS, a generally accepted value in activated sludge literature. In one set of experiments, mixed liquor COD analyses confirmed that this value was a reasonable estimate of the biomass COD content. The difference between the predicted carbonaceous oxygen demand based on the COD balance and the observed carbonaceous oxygen demand based on corrections to the measured oxygen uptake rates was attributed to a COD loss, and was attributed to the use of an anaerobic contact zone in an A/O system. This loss was referred to as anaerobic stabilization.

Table 4-1 summarizes the percent of COD loss observed for different experiments reported by Randall et al (1992). All of the studies were done in small systems; three in bench-scale reactors with reactor volumes in the range of 1.0–4.0 L and one in a pilot plant study with individual reactors of approximately 50 L (13 gallons). Studies 3 and 4 involved experiments with domestic wastewater and the other two used synthetic wastewater. Study 4 was the only one in which a parallel aerobic-only system was operated without an anaerobic contact zone. The following observations can be made from the results presented:

- For synthetic wastewater the COD loss reported was significant, ranging from 23–48%. When dextrose (25% of feed COD) was replaced with acetate, the COD loss calculated was 10% or less.
- For Study 3, the COD loss was found only when the influent COD was greater than about 210 mg/L. The authors claimed that anaerobic stabilization (COD loss) occurred only for the COD in excess of that needed for biological phosphorus removal. In this case the wastewater was phosphorus-limited and excess COD was available. It should be noted that most municipal wastewaters are not phosphorus-limited. The authors further claimed that the COD loss likely was due to non-phosphorus accumulating organisms (non-PAOs).
- The COD loss in Studies 1 and 2 represented 10–48% of the COD removal in the anaerobic step.
- For Study 4, which had an aerobic-only process as a control versus an EBPR UCT process, the COD loss observations based on the COD balance and oxygen uptake results did not agree with the observed MLVSS concentration comparisons. The SRTs of the UCT process and the conventional aerobic process were reported as equal along with equal reactor aerobic volumes and feed flows. With identical SRTs, reactor volumes, and loadings, a lower MLVSS would be expected for the UCT process with the reported COD loss, as less COD would be available for biomass production. However, as summarized in Table 4-2, the UCT system in most cases had a higher MLVSS concentration. This result is contrary to expectations for solid production based on the COD loss calculation.

Table 4-1. Summary of Test Results for Reported COD Loss in A/O Processes (Data from Randall et al., 1992).

Study	Type	Process	Feed	SRT, days*	Total % Volume Anaerobic	COD Loss, %	COD loss as % of Anaerobic COD uptake
1	Bench	A/O	70% Nutrient Broth 25% Dextrose	5-10	30	23-48	13-48
2	Bench	UCT	68% Nutrient Broth 27% Dextrose	12-18	32	23-27	10-21
			58% Nutrient Broth 37% Acetate	13	32	0-10	0-12
3	Pilot Plant	UCT	Domestic Raw Wastewater	5.3-10	21-28	0-26	N/A
4	Bench	UCT	Municipal Wastewater	5-15**	25	8-27	N/A

* When range given, tests were done at different SRTs.
** Temperature was varied at 10, 15, and 20 °C.
N/A=Not available

Table 4-2. Comparison of Anaerobic COD Loss to Reactor MLVSS Concentrations Between EBPR Process (UCT) and Aerobic Only Process (Data from Randall et al., 1992).

Phase	Total SRT, days	Temp., ^{0}C	Reported% COD Loss*	UCT MLVSS, mg/L	Aerobic Only MLVSS, mg/L	UCT to Aerobic VSS ratio
IA	5	10	18.3	1180	1370	0.86
IB	5	15	8.4	1080	1050	1.03
IC	5	20	10.0	1420	970	1.46
IIA	15	10	11.6	2290	2090	1.10
IIB	15	15	24.6	2650	2100	1.26
IIC	15	20	27.3	2300	1270	1.35

Barker and Dold (1995) applied the same COD mass balance technique, as described for Randall et al. (1992), with data from bench-scale studies from a number of sources that used either synthetic wastewater or municipal wastewaters. The authors stated that the data used for their mass balances were obtained from steady state operating periods, but the number of data points used was not given. A unique feature of their evaluations was that they compared the COD balance recovery to a nitrogen balance recovery. The nitrogen balance considered the net fate of nitrogen across the systems. The mass of nitrogen into the system was expected to equal the nitrogen in the effluent stream as the sum of organic, ammonia, and oxidized nitrogen, plus the amount of nitrogen in the waste solids, plus the amount of oxidized nitrogen reduced by denitrification. The benefit of including the nitrogen balance is that a COD loss calculation error due to inaccuracies in the values for nitrogen oxidation and/or reduction would be revealed by a poor recovery in the nitrogen balance. As indicated in Table 4-3, the authors achieved almost full recovery of nitrogen for all systems except the anoxic/aerobic process, which had a nitrogen balance that was 10% higher than expected.

Table 4-3. Summary of Nitrogen and COD Balance Recoveries for Bench Scale Activated Sludge Processes with and without an Anaerobic Contact Reactor (Data from Barker and Dold, 1995).

Referred Study*	Process	Substrate	N. Balance Recovery	COD Balance Recovery
McClintock	Aerobic	Synthetic	99.4%	94.2%
McClintock	Anoxic/Aerobic	Synthetic	110.4%	96.9%
Arkley & Marais	A/O	Domestic WWT	99.8%	76.5%
Wentzel et al.	UCT	Domestic WWT	101.2%	75.2%
Wentzel et al.	MUCT	Domestic WWT	97.1%	79.2%
Wentzel et al.	Bardenpho and UCT	Acetate	98.0%	91.0%

*Referenced Studies in Barker and Dold (1995).

In comparison, the authors found poor COD recovery in their balances for systems with an anaerobic contacting zone as used in an anaerobic/aerobic (A/O) process, a University of Capetown (UCT) process, and a Modified University of Capetown (MUCT) process fed domestic wastewater. For these systems the mass balance showed a COD loss of 21–25%, in a similar range to that reported by Randall et al. (1992). For systems without the anaerobic contacting zone (the aerobic and anoxic/aerobic system fed synthetic wastewater), the COD balance was comparable to the nitrogen balance and no significant COD loss could be claimed.

In a study by Wentzel and co-workers referenced by Barker and Dold (1995), four separate systems incorporating an anaerobic contact zone were operated with only acetate as the carbon substrate to study biological phosphorus removal. These results are averaged in the last row of Table 4-3 and are for three studies with the Bardenpho Process at SRTs ranging from 7.5–20 days and one UCT Process at an SRT of 10 days. In this case the COD recovery ranged from 89–92% and on average was only 9% less than expected.

4.2.2 Possible COD Loss Mechanisms

Both Randall et al. (1992) and Barker and Dold (1995) associate the COD loss with the use of an anaerobic contact zone. Randall et al. (1992) also referred to the loss as an anaerobic stabilization. Considering the anaerobic conditions in this zone and the type of biological reactions possible for an anaerobic environment, the possible COD loss mechanisms are presented and discussed in the following:

- Hydrogen production and stripping in the anaerobic contact zone.
- Sulfate reduction with COD consumption in the anaerobic contact zone.
- COD utilization by oxygen-tolerant methanogens in the anaerobic zone.
- Production of volatile fatty acids in the anaerobic zone and subsequent stripping in the anaerobic and/or aerated zone(s).
- Error in the measurement of COD, oxygen uptake rate, or solids produced.
- Effect of surface transfer of oxygen in the anaerobic zone.

Research by Wable and Randall (1994) was aimed specifically at determining the mechanism for the COD loss. Bench-scale reactors fed synthetic wastewater were operated with an anaerobic contacting zone to encourage EBPR. Two systems were tested, an A/O system and an A^2O system. Using oxygen uptake rates and the COD balance technique, significant levels of

COD loss were observed; an average of 15% for the A/O system and 45% for the A^2O system. Tests were done to determine if the loss could be explained by evolution of fermented products from the anaerobic zone, such as hydrogen stripping or volatile fatty acids, but none were detected. However, the analytical technique for measuring hydrogen in the gas phase was gas chromatography with a thermal conductivity detector, which has limited sensitivity.

1. Hydrogen production and stripping in the anaerobic contact zone.
A possible mechanism for COD loss, presented by Barker and Dold (1995) and Randall et al. (1992), is hydrogen production and evolution. Hydrogen is a known product of anaerobic fermentation; for anaerobic sludge digestion, about one-third of the methane produced is estimated to be via hydrogen production from the transformation of particulate degradable COD (Tchobanoglous et al., 2003). However, not all fermentation reactions produce hydrogen and the type of fermentation bacteria and reactions in EBPR anaerobic contact zones is unknown. Hydrogen is a poorly soluble gas and for small-scale reactors with shallow depths and relatively high surface/gas interfacial area relative to the liquid volume, hydrogen could be stripped to the atmosphere. In full-scale systems with much lower surface areas to volume ratios, significant stripping is less likely. Another pathway for hydrogen loss is consumption by methanogens and methane stripping. Wable and Randall (1994) used a vacuum stripping technique to obtain gas samples from the mixed liquor in the anaerobic contact zone for analyses for hydrogen and methane on a gas chromatograph with a thermal conductivity detector. The detection limit of the hydrogen measurement was stated as 0.001 mg/L, which may not have been sensitive enough to measure the low level of hydrogen necessary to allow fermentation with sustained hydrogen production. Thus, even though they did not detect hydrogen in the reactor, the measurement method and operation condition resulted in an inconclusive finding. Hydrogen may have been present and not detected, or produced and lost from the mixed liquor via transport through the surface due to the relatively large surface area to volume and the poor solubility of hydrogen.

Barker and Dold's (1995) findings strongly support the possibility of COD loss via the fermentation process and production and stripping of hydrogen. The results in Table 4-3 show that little COD loss occurred when acetate was the organic substrate fed to the anaerobic contact zone versus acetate production for EBPR by fermentation of domestic or synthetic substrates. It should be noted that these tests were not done with the intention of studying COD loss and the test conditions occurred at different times. Without fermentation there is no chance of hydrogen production and loss. The comparison of the COD loss with a more complex feed versus acetate suggests that fermentation may have played an important role in COD loss.

2. Sulfate reduction with COD consumption in the anaerobic contact zone.
Under anaerobic conditions, sulfate can serve as an electron acceptor. The oxygen equivalent of sulfate is 0.67 g 0_2 /g SO_4 and an average sulfate concentration for domestic wastewater is 30 mg/L (Tchobanoglous et al., 2003). Thus, assuming a synthesis yield of 0.60 g biomass COD/g COD used, the amount of influent COD that could theoretically be accounted for by sulfate reduction is about 50 mg/L. This could account for about 25% of the biodegradable COD of a typical domestic primary effluent, so that sulfate reduction provides a possible avenue for COD loss that was not accounted for in the method used for the COD balance. On the other hand, sulfate reduction may not be possible in the sequential A/O processes, as oxygen is expected to inhibit sulfate-reducing bacteria so that they may not be active in the anaerobic contact zone.

Studies on COD loss have not evaluated this pathway, but its role should be investigated in further experiments on COD loss.

3. COD utilization by oxygen-tolerant methanogens in the anaerobic zone.
Theoretically methanogenesis could explain the COD loss based on the COD balance method used. Barker and Dold (1995) point out that oxygen-tolerant methanogens exist but there has been no work showing that they can grow in an environment with a majority of their time under oxygen exposure. Methanogens are known for their relatively slow growth rates. Using typical growth kinetics reported for methanogens at 25 °C (Tchobanoglous et al., 2003), the minimum SRT needed to maintain them in a system is about six days. Considering that the anaerobic contact zone represents about 10–20% of the total system volume, the total SRT for a EBPR system with COD loss via methanogenesis in the anaerobic zone would have to be in the range of 25–50 days at 25 °C. Thus, COD loss by methanogenesis seems unlikely. Wable and Randall (1994) were not able to measure any methane production from mixed liquor samples in the anaerobic contact zone of their laboratory A/O and A^2/O systems when fed a bactopeptone or acetate feed. The ability of slow growing, acetate-utilizing methanogens to compete with PAOs for acetate would also be very unlikely considering the inability to measure any residual acetate in the anaerobic zone of EBPR systems.

4. Production of volatile fatty acids in the anaerobic zone and subsequent stripping in the anaerobic and/or aerated zone(s).
Barker and Dold (1995) list this as a possible pathway for COD loss, but then dismiss it in view of the rapid uptake of such substrates relative to the possible low stripping rate associated with their high solubility and relatively low Henry's constant.

5. Error in the measurement of COD, oxygen uptake rate, or solids produced.
The method used to assess COD loss was highly dependent on the oxygen uptake rate (OUR) measurements and COD measurements, as well as proper influent and effluent sampling. For some reported experiments *ex situ* OURs were done, which could give lower values and subsequently a calculated COD loss. However, other reported tests with COD loss did use *in situ* measurements. It is difficult to fully assess the effect of experimental variability and potential errors in the COD loss reports because no detailed accounting was given on the number of data points used and variability of solids measurements and OURs.

The COD content of the wasted solids is an important factor in the COD loss calculation. Randall et al. (1992) claimed test data for which the COD content of the waste biomass agreed with the theoretical value of about 1.4 g COD/g VSS. If the COD content was actually higher, an erroneous COD loss would be calculated. Wable and Randall (1994) raised some concern about the inability of the COD test to accurately measure the COD content of waste biomass by the possible inability to chemically oxidize all of the cell constituents. Through separate tests on NAD and NADH chemical oxidation, they showed that these compounds could not be easily oxidized. They further point out without verification, that if NAD and NADH were at a higher level in PAOs in EBPR systems, a COD loss would be erroneously calculated. The significance of these observations is that the COD mass balance calculation used to determine a COD loss has relied on a test method that does not always provide 100% oxidation.

6. Effect of surface transfer of oxygen in the anaerobic zone.

None of the previous investigators have addressed the potential issues related to surface transfer of oxygen in the anaerobic zone of the experimental systems used to study COD loss in EBPR systems. All of the test systems were relatively small reactors with a high surface area to volume ratio as compared to tanks typically found in full-scale systems. With reactor depths typically only a few inches, compared to depths of 12–16 ft for full-scale EBPR systems, the test systems had surface area to volume ratios about 20-30 times that of full-scale applications. This geometry can allow for significant transfer of constituents into or out of the anaerobic contact zone that may effect the interpretation of a COD loss for the system.

If an anaerobic atmosphere is not maintained above the reactor liquid, significant oxygen transfer may occur, which would be part of a COD loss not identified by the COD mass balance method. Inefficient biological phosphorus removal can occur in small reactors due to oxygen transfer into the reactor surface. Nitrogen sparging has been used in such reactors to prevent COD consumption by oxygen provided from surface transfer (Stephens and Stensel, 1998).

4.3 Reported Process Performance for Biomass Reduction

COD loss has only been studied in bench or small pilot-scale A/O processes. As most of these studies have been done with single systems and without a parallel control system without the anaerobic step, there is no data available reporting actual biomass production with and without the anaerobic step. The COD loss has been reported as high as 48% in Table 4-1 and in the range of 8–27% in Table 4-2.

In one case, data has been provided to assess the solids yield for an A/O system operation (Randall et al., 1992) fed synthetic wastewater. Based on the data provided, the observed yield ranged from 0.16 to 0.22 g VSS/g COD removal for SRTs in the range of 4.5–10.3 days. The following model, describing net biomass growth as a function of substrate utilization and endogenous decay, was used to determine the synthesis yield.

$$\frac{1}{SRT} = u_{net} = Y(U) - K_d \qquad 4.3\text{-}1$$

Where:

SRT = Solids retention time, days

u_{net} = net specific biomass growth rate, g VSS/g VSS-d

Y = Biomass synthesis yield coefficient, g VSS/g COD

U = Specific substrate utilization rate, g COD/g VSS-d

K_d = endogenous decay coefficient, g VSS/g VSS-d

A fit of the data provided shows a synthesis yield coefficient value of 0.37 g VSS/g COD removal and endogenous decay value of 0.06 g/g-d (Figure 4-6). The synthesis yield coefficient value is within the range of values normally reported for aerobic treatment of organic substrates (Tchobanoglous et al., 2003.).

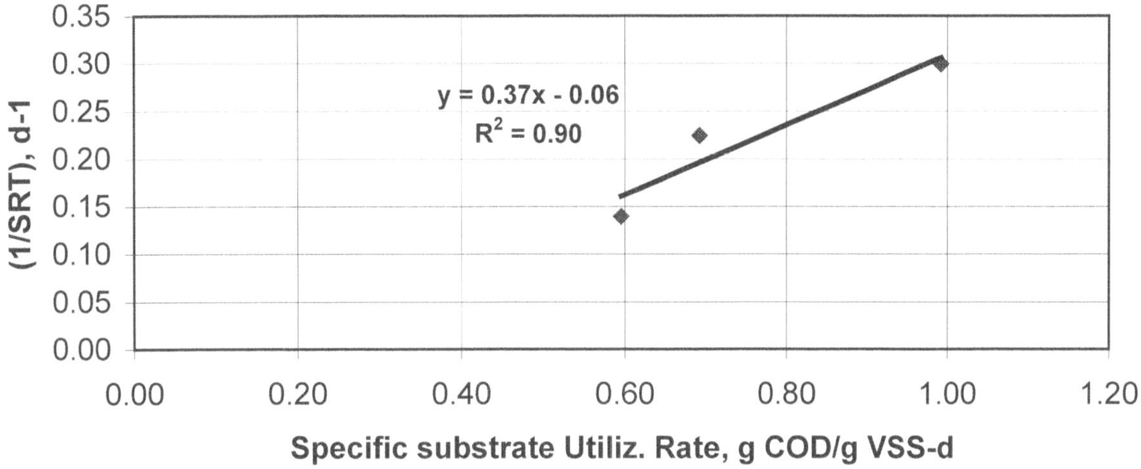

Figure 4-6. Solids Yield Evaluation for Anaerobic/Aerobic Process Treating Synthetic Wastewater (Randall et al., 1992.)

The COD loss function has been incorporated into the various versions of the BioWin™ model (Barker and Dold, 1997). For configuration with an anaerobic contact zone, the model predicts a COD loss and subsequently a lower oxygen demand in the aeration reactors and less biomass production. The earlier version assumed a default value of about 80% COD loss associated with fermentation. The current version uses a different approach based on hydrogen evolution during fermentation of soluble readily biodegradable substrates. It assumes that 35% of the COD is converted to hydrogen and about half of the hydrogen is lost, equating to a COD loss of about 17% of the rbCOD. Subsequently, the biomass production would be less by 17% times the fraction of influent COD that is biodegradable. A 30–50% fraction is possible so the reduction in biomass production predicted would be on the order of 5–9%. This is less than that indicated by Randall et al. (1992) and Barker and Dold (1995). Based on a 30–50% rbCOD fraction and their observation of 20–30% COD loss, the estimated reduction in biomass production would be 6–15%.

In spite of the promising results for a COD loss in prior studies, researchers have focused on reduced oxygen needs rather than reductions in sludge or biomass. No data have been reported showing differences in biomass or VSS production for systems with and without the anaerobic contact process used in EBPR systems.

In the bench-scale tests conducted as part of this project (Chapter 10.0), no reduction in VSS production was found for studies with anaerobic/aerobic and aerobic systems operated in parallel with the same feed and operating SRT. In cases where OUR data were available, COD balances were done, but the balances did not indicate any COD loss.

4.4 Plant Retrofit Considerations

The addition of an anaerobic contact process is a relatively simple change to activated sludge processes operated at relatively low SRTs for only BOD removal. A typical hydraulic detention time for the contact basin is one hour and modest mixing (about 0.01 kw/m^3) needs are provided for the tank (Randall et al., 1992). The additional tankage requirements become more significant where nitrification is occurring in the activated sludge process. As shown in Section 4.1, tankage for nitrate removal would also be required so that the total tank volume is sufficient to provide an additional HRT of about 2-3 hours. However, it has been shown that some or all of this volume may be obtained from the existing aeration volume due to a possible reduction in the needed aerobic volume because of the ability to carry a higher MLSS concentration. Higher MLSS concentrations are made possible by the greatly improved solids thickening characteristics associated with anaerobic contacting selector processes (Albertson, 1987).

4.5 Assessment of Process

The main advantages of the anaerobic/aerobic EBPR process are the ease of implementation, improved operations due to sludge volume index (SVI) control, and the ability to remove nutrients. Because of the improved SVI and ability to carry higher mixed liquor concentrations, the need for additional tankage may be minimal or non-existent and would have to be considered on a site-by-site basis.

At present, the main incentives for using anaerobic contact processes in activated sludge treatment are the improved sludge thickening characteristics (SVI control) and the need for nutrient removal. If it can be shown that less sludge production is possible, a further incentive would be found for implementing various EBPR processes. However, with lack of understanding of the mechanism for COD loss and the uncertainty of its existence based on this work, claims of reduced biomass production may not be justified. In fact, the research results reported here suggest that the COD losses observed previously could be the result of an artifact of testing in small-scale systems with surface transfer of oxygen or of problems in the COD test method unique to biomass produced in EBPR systems. Without more definitive proof for COD loss, EBPR systems with A/O treatment steps can not be recommended for use on the basis of reduced biomass production.

4.6 Research Needs

The results from this research (Chapter 10.0) suggest that the COD loss mechanism and reduced biomass production may not be realized, based on volatile solids production measurements between EBPR and conventional aeration processes. Before the COD loss can be claimed and credit taken for reduced solids production, a fundamental and creditable hypothesis for the COD loss mechanism needs to be developed and tested. The research results and more in-depth review of possible COD loss mechanisms do not support any of the current mechanisms proposed. In fact a fundamental evaluation of the most promising mechanism put forth, the production and stripping of hydrogen, shows that this pathway is not likely. Further evidence of this is the reported levels of propionate in fermenters used to supplement the rbCOD to EBPR systems. Because of the difficulty of measuring 10–20% differences in solids production, more innovative methods are needed in further research to evaluate biomass reduction in EBPR processes. The method needs to measure a parameter related to biomass growth and to have a measurement accuracy greater than for the VSS test. Future investigations should consider more sophisticated analytical tools to follow the fate of substrates in the anaerobic/aerobic and aerobic processes to better evaluate biomass production. Without more direct proof of COD loss, designers may be over-optimistic if they design EBPR facilities assuming reduced oxygen transfer needs and less sludge production due to COD loss.

CHAPTER 5.0

CELL DISRUPTION PROCESSES BY MECHANICAL AND THERMAL METHODS

5.1 Description of Processes

In mechanical or thermal cell disruption processes, physical forces or heat is applied to the activated sludge biomass to a sufficient degree to cause the bacteria cells to rupture and release intracellular components, which include protoplasm, proteins, and nucleic acids (Foster, 1992). In essence the disruption process is the equivalent of increasing the biomass endogenous decay rate. Thus, for a given activated sludge process SRT a greater amount of cell lysis occurs due to the mechanical or thermal process applied, which will then decrease the net biomass yield and total solids production. A higher oxygen consumption rate would be needed for biological oxidation of the released biodegradable substrates, which would require additional oxygenation capacity for activated sludge processes.

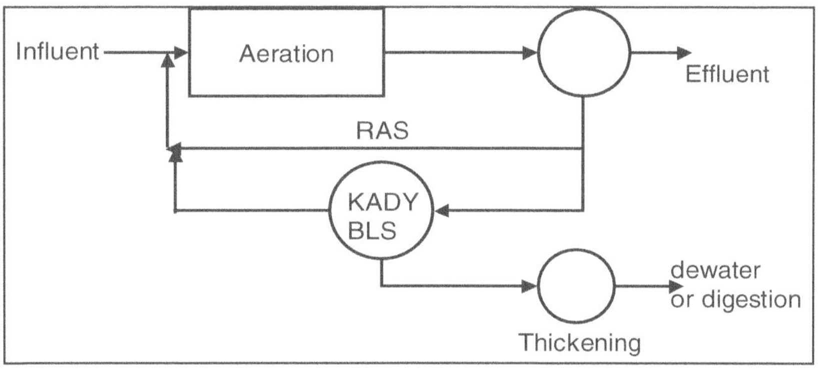

Figure 5-1. Schematic of Mechanical Cell Disruption Process Using Kady Biolysis System.

Cell disruption processes are typically applied at the RAS recycle stream. One such system that has been tested in pilot and bench-scale studies for municipal activated sludge treatment is the KADY Bio-Lysis process (Springer et al., 1996; Wadehra et al., 1999), which is illustrated in Figure 5-1. In this system a portion of the RAS flow is processed through a high speed mechanical shear device that disrupts the floc and biomass cells and possibly releases intracellular materials to solution. The KADY mill consists of a slotted rotor inside a slotted stationary cylinder of a few thousands of an inch. The rotor turns at high speed to force material through the radial slots. The velocity is estimated at about 2745 m/min (9000 ft/min) (Springer et al., 1996). The high speed impact of the sludge hitting the stationary rotor is thought to be sufficient to rupture biological cells. About 3–5% of the activated sludge inventory is processed through the mill machine daily, and the energy consumption of the machine is about 6.9 kw/1000m^3 (35 Hp/Mgal/day) of wastewater flow treated. Typical machine unit sizes used, as reported in full-scale testing, are in the 7.5–19 kW (10-25 Hp) range.

Other mechanized methods have been used to cause bacteria cell disruption. Discharging activated sludge through high pressure jet nozzles with pressure drops exceeding 5000 PSI can cause cell disruption (Foster, 1992). A U.S. patent by Stephenson and Dhalival (1998) describes a jet nozzle cell disruption system. In their description, the authors site the preferred application as having the pH raised to 11.0 by NaOH and mechanical maceration to reduce the particle size before pumping through the high pressure nozzle. They proposed using this method prior to sludge digestion versus using it in a liquid treatment process. Similarly, ultrasonic treatment has also been proposed as a sludge pretreatment step to sludge digestion, with the benefit of increasing digestion rates. Tiehm et al. (2001) claimed that ultrasonic pretreatment can significantly reduce aerobic or anaerobic sludge digestion times for the same degree of volatile solids reduction without ultrasonic treatment.

A thermal treatment process, which is incorporated into an activated sludge system in the same way as for the mechanical system show in Figure 5-1, has been described by Sakai et al. (2000). A thermal treatment reactor is used in place of the KADY Bio-Lysis reactor in Figure 5-1. A portion of the RAS is held in the thermal treatment reactor for 24 hours at 65 °C, which is inoculated with *Bacillus stearothermophilus*, a thermophilic bacteria with high proteinase production, to accelerate biomass destruction. In another thermal treatment method described by Rocher et al. (1999), high temperature treatment was coupled with pH adjustment in tests with high pH alkaline conditions (pH 10.5–11.5) and acid conditions (pH 1.0–3.0). The best results were found at a temperature of 100 °C and with high pH.

5.2 Process Fundamentals and Biomass Reduction Mechanisms

The ability to rupture cells in mechanical systems, such as high pressure homogenizers or ball mills, is believed to be due to cavitation caused by the high speed rotating equipment (Foster, 1992). The cavitation causes an ultrasonic resonance, which in turn leads to a vibration that ruptures cells.

A similar explanation is given for ultrasonic treatment by Tiehm et al. (2001). The ultrasonic waves cause periodical compression and rarefaction of the medium leading to cavitation. Cavitation creates gas bubbles, which first grow in size and then collapse within a few microseconds. The rapid collapse creates violent hydromechanical shear forces in the bulk liquid which can disrupt bacteria cells. The extreme conditions have also been associated with some thermal destruction effects and the creation of very reactive hydroxyl radicals. The sonification process was reported to be more effective at a frequency above 100 kHz.

High temperature causes both deactivation of bacteria cells and lysis with release of cellular material. Rocher et al. (1999) found that as the temperature was raised from 60 to 100

°C, the amount of dissolved organics released from a pure culture of *Alcaligenes eutrophus* increased from 28–66% during a one-hour incubation period. In the process of high temperature and bioaugmentation with thermophilic bacteria (Sakai et al., 2001), the bacteria were thought to cause greater solubilization of cell material than by temperature alone, due to the activity of proteinase enzymes.

5.3 Reported Performance

***KADY Biolysis Process**:* One of the first reports on performance with the KADY process was by Springer et al. (1996), based on laboratory batch tests. They reported volatile solids conversions of 8–10% for operation at ambient temperatures versus 55–76% conversion under uncooled mill temperatures of 40 °C. The process has also been evaluated at full-scale and pilot-scale operations. In many cases the solids production was determined for plant operations before and after the KADY mill installation. However, without concurrent parallel operations, with and without RAS treatment with the KADY mill, it was difficult to determine whether solids production differences were due to changes in the wastewater characteristics, changes in the process operation, or to the KADY process installation.

Solids reduction for the KADY process versus that for a control activated sludge operated in parallel without mechanical disintegration has been reported for three cases: Portland, ME; Detroit, MI; and Springfield, MA. Though the solids production was lower by over 50% in a treatment train with the KADY process versus a train with the normal operation (Springer, 1999) at Portland, the results are inconclusive because the SRT was higher by a factor of 3-4 for the treatment train with the KADY process. Thus a good portion, if not all, of the solids reduction may have been due to the higher SRT operation. For the Detroit study (Wadehra et al., 1999) a dual-pilot system was operated in parallel at a nominal 37.9 L/min (10 gpm) flow rate with SRT control. The aeration tank volume was 1350 gallons, resulting in a 2.25-hr HRT. The clarifier solids were included in the sludge inventory calculation for SRT control. Return sludge was batch fed to a holding tank each day prior to processing by the KADY milling equipment. The sludge was withdrawn from the holding tank, and processed through the KADY machine, and fed back to the activated sludge system on a continuous basis over a 16–20 hour period.

The results for the three operating phases for the Detroit study are summarized in Table 5.1. The amount of activated sludge inventory milled in phases I, II, and III were estimated at 3–5%, 8–10%, and 8–10% respectively. The waste solids reduction was reported to be significant with the KADY operation, with 27–42% less sludge production compared to the conventional process for the three operating phases at SRTs ranging from 7.3 to 21.9 days. The effluent quality or sludge settling characteristics were not significantly affected by using the KADY process. For phases I and II, the average SVI was slightly less for the KADY process and about half as many filamentous bacteria were observed under a phase-contrast light microscope.

Table 5-1. Average Performance Reported for KADY Treated Versus a Control Activated Sludge Train (Wadehra et al., 1999).

Parameter	Unit	*Phase I* KADY	CTRL	*Phase II* KADY	CTRL	*Phase III* KADY	CTRL
Flow	L/min	33.5	31.8	33.2	31.0	34.2	32.2
PE $cBOD_5$	mg/L	47	45	65	49	115	93
PE COD	mg/L	164	155	214	184	321	281
PE TSS	mg/L	54	52	73	65	72	70
PE TKN	mg/L	10.7	10.8	12.7	12.4	18.9	18.1
SRT	days	7.3	5.4	21.9	20.0	20.4	14.6
MLSS	mg/L	2,535	2,060	5,100	5,585	4,646	4,541
RAS	mg/L	10,445	9,004	14,652	14,658	13,490	12,937
SVI	mL/g	86	103	86	101	114	89
SE COD	mg/L	45	49	63	68	47	53
SE $cBOD_5$	mg/L	17	18	9	10	18	16
SE TSS	mg/L	13	20	23	28	16	27
SE TKN	mg/L	7.0	7.2	2.9	3.3	4.9	4.4
Sludge Prod.	kg/d	2.47	2.84	2.00	2.46	2.27	2.63
AS Reduction	%	26.5	—	41.5	—	32.4	—

PE = primary effluent; SE = secondary effluent; AS = activated sludge

The results for the Detroit tests are further evaluated in Table 5-2, in which the observed yields in terms of gVSS/g BODr are compared between the trains, along with their difference and the theoretical observed biomass yield. The theoretical observed biomass yield is based on the following equation, which is a component of Equation 1.1-2. Typical synthesis yield (0.60 g VSS/g BOD_r) and endogenous decay coefficient (0.08 g VSS/g VSS –d) values (Tchobanolgous et al., 2003) were applied.

$$Y_{obs,\ biomass} = \frac{Y}{(1+K_d SRT)} \qquad 5.3\text{-}1$$

Where:

$Y_{obs,\ biomass}$ = observed biomass yield, g VSS/g BOD_r

K_d = endogenous decay coefficient, g VSS/g VSS –d

Y = synthesis yield, g VSS /g BOD_r

Table 5-2. Evaluation of Observed Yields for a Parallel Two-Train Pilot Plant Study with KADY Mill Treatment of RAS for One Train (Wadehra et al., 1999).

Phase	Average SRT, d	*KADY Y_{obs},* g VSS/g BOD_r	Control Y_{obs}, g VSS/g BOD_r	Difference (Control minus KADY Y_{obs}), g VSS/g BOD_r	Est. $Y_{obs,\ biomass}$*, g VSS/g BOD_r
I	6.4	1.14	1.64	0.50	0.40
II	21.0	0.65	0.99	0.34	0.22
III	17.5	0.36	0.55	0.19	0.25

*Based on Equation 5.3-1.

The differences between the observed yields for the control and KADY trains ranged from 0.19 to 0.50 g VSS/g BOD_r, which follows the respective solids reduction efficiencies reported for each phase. Because the basis for solids reduction is rupturing of biological cells, these differences should represent some fraction of the biomass produced. However, a comparison of the Y_{obs} destroyed (i.e., KADY Y_{obs} minus the control Y_{obs}) to the theoretical biomass yield shows that the amount of biomass destroyed is greater than the theoretical biomass production for phases I and II and 78% of the biomass produced in Phase III. The inability to develop a reasonable comparison between the theoretical biomass production and solids reduction suggests some chronic problem in the pilot plant operation and in data collection, such that the results must be deemed inconclusive.

A KADY Biolysis system was installed in two trains of a four train activated sludge system at Springfield, MA (Camp Dresser and McKee, 2000), and the two parallel systems were evaluated for comparative treatment performance and observed sludge production. The parallel systems were operated for about 10 months from May 1999 to March 2000. Each of the two trains received about 20 Mgal/d of flow during the study period. The aeration systems were preceded by primary treatment and were operated at SRTs in the range 18–27 days. The KADY trains had an average observed solids yield of 0.84 g TSS/g BOD_r versus 0.96 g TSS/g BOD_r for the control system, which is a 12% lower solids production. However, the operating SRTs were not equal between the trains, with higher SRTs in the KADY train. A graphical evaluation of the observed yield versus SRT indicated a higher observed yield for the KADY system at a given SRT by 10–15%. The report also notes that it was impossible to assure that the flow split between the two systems was equal, as no online flow measurement equipment was available for each train. Thus, the results of this study are inconclusive with regard to the solids reduction ability of the KADY process.

Other observations for the Springfield study were a higher SVI for the KADY train (138–150 mL/g versus 71–78 mL/g for the control side), a lower average effluent TSS and BOD concentration for the KADY trains by about 2 mg/L, and a higher oxygen transfer efficiency for the KADY train in one case, but relatively equal in another case.

It should be noted that the Springfield facility has opted to reduce its solids production by operating at a higher SRT. Facility operators found that they could raise the aeration tank MLSS concentration from 2,000 to 5,000 mg/L, which increased the SRT from 4 days to 10 days. The higher SRT produced a 30% reduction in solids production (24 dry tons/day to 17 dry tons/day). The reported savings for the sludge reduction using the higher SRT operation was $780,000/year or about $305/dry ton. The operators reported an increase in power costs of only $50,000 per year, leaving a net savings of $730,000/year.

***High Pressure Homogenizer**:* Camacho et al. (2002) reported on bench-scale tests with a high pressure homogenizer (HPH). Two parallel 200-L (52.8 gallon) activated sludge aeration tanks with membrane filtration for liquid-solids separation were operated in parallel, with the mixed liquor of one of the reactors recycled through the mechanical disintegration equipment. The side-by-side reactors were operated at an SRT of 10–15 days with two passes of HPH treatment for the one reactor during one phase of the study. The mechanical device had an operating pressure that ranged from 300–700 bars. The test variable was the number of times the sludge removal from the aeration tank was cycled through the HPH. In all cases, 20% of the reactor mixed liquor volume was treated daily. The degree of solubilization was assessed by analyzing the COD concentration of HPH effluent samples after $FeCl_3$ coagulation and centrifugation. Released COD was judged by unsettleable solids.

Camacho et al. (2002) found that they could achieve up to 80% COD release at 300 or 700 bars, but only 18–25% COD solubilization as the number of passes was increased from one to seven. At one pass, the COD solubilization was less than 10%. The HPH treatment was found to reduce the biomass production by 19% with the observed yield decreasing from 0.36 to 0.29 g TSS/g COD_r. The control reactor had a higher SVI (336 mL/g versus 173 mL/g) and lower effluent TSS concentration (4 mg/L versus 18 mg/L) than the HPH treated reactor.

***Thermal Treatment**:* Sakai et al. (2000) reported on the results of a 400 m^3/d (73 gpm) pilot plant study with thermal treatment of a portion of the sludge recycle stream. Thermophilic bacteria with high proteinase ability were added to the thermal reactor that operated at 65 °C for 24 hours. The influent to the activated sludge process was municipal wastewater with average BOD_5 and TSS concentrations of 125 and 100 mg/L, respectively. The treated effluent BOD_5, TSS, and NH_4-N concentrations were about 10, 10, and 12 mg/L respectively. The high temperature RAS treatment apparently inhibited the nitrification efficiency. The reported observed yield was 0.15 g TSS/g BOD_r, which is about 25% of what would be expected for conventional activated sludge treatment.

Rocher et al. (1999) compared the dissolved organic carbon (DOC) released from waste activated sludge in batch high temperature tests at 100 °C with varying pH to that achieved by sonification. With an incubation time of 45 minutes and acid or base added to vary the pH values from 4.5 to 8.0, the DOC release was only 25–50% of that from sonification. At a pH of 11.5 and 12.0 the DOC release was 125% and 220%, respectively, of sonification. It should be noted that sonification was done only at 20 KHz, which may not be at an optimal level. The authors further showed that 80-90% of the released DOC could be biodegraded. At the higher pH of 12, the DOC released was 350 mg/g TSS at 100 °C, compared to 100 mg/g TSS at a 7.0 pH. Assuming 85% volatile solids for the biomass and 50% carbon, the available carbon is approximately 425 mg/g TSS. Thus the percent of the cell carbon released ranged from 24–82% from pH 7 to pH 12 at 100 °C. The authors claimed that high pH also has been shown to enhance biomass disintegration in a sonification and a HPH treatment.

5.4 Plant Retrofit Considerations

All of the sludge reduction methods described in this chapter involved treating a side stream of the RAS flow, which should be relatively easy to implement in a plant retrofit provided that space is available. For the KADY process and the HPH application, no detrimental effect on effluent quality was found. It was interesting to note that the mixed liquor sludge settling and thickening characteristics were different between the mechanical treated and conventional processes as indicated by the SVI test results, and likely explain the differences observed in effluent TSS concentrations. A lower TSS was observed for cases with a higher SVI. When higher SVIs are associated with a higher level of filamentous bacteria, a lower TSS concentration is generally found as the filamentous structure tends to trap smaller particles.

Installation of a high temperature contact unit for RAS treatment appears more problematic than the installation of a machine to process sludge as in the KADY and HPH processes. The high temperature treatment would require the installation of a 24-hour contact unit, which would require a significant amount of space. In addition, heat exchange equipment and an energy source would be needed to raise the return sludge temperature to 65–100 °C, depending on the process. For the application with bio-augmentation with thermophilic bacteria, the process installation would be further complicated by the possible need to have storage and maintenance of innocule.

5.5 Assessment of the Mechanical and Thermal Treatment Processes

Mechanical disintegration processes. The feasibility of installing mechanical processes on the RAS line of an activated sludge process has been demonstrated via full-scale testing with the KADY process. In addition, no secondary problems—such as upsetting the plant operation or negatively affecting plant performance—have been indicated in full-scale tests. Additional oxygen transfer capacity would have to be added or available for oxidation of released substrate from the cell disruption processes.

An operational issue related to the installation of mechanical devices is the solids pretreatment before the mechanical equipment and the equipment maintenance. On all the full-scale tests KADY personnel operated the equipment, which required pre-screening of the solids. Thus, the process does add an additional and significant burden on plant operations.

Another practical and economic issue for the installation of the KADY process is the number of units needed, considering that their largest reported unit was 25 Hp. Following reports of major solids reduction (26-42%) in the Detroit pilot plant study, the process surprisingly was not tested at full scale or considered for full-scale installation. One reason given to this author was that the large number of KADY units that would have to be installed and maintained raised significant concerns about operational requirements.

The major issue with the mechanical disintegration processes is the degree of solids reduction possible and proven, and whether a net economic benefit exists when the trade-off of reduced sludge processing costs are compared to increased costs for the mechanical equipment, operations and maintenance, and power costs. The first part of any economic evaluation would have to be the degree of solids reduction made possible by installing and operating the mechanical treatment device.

For the KADY process, the test results are inconclusive with regard to solids reduction. A well-controlled certifiable test is still lacking. The Springfield tests may be the most reliable in view of having a large full-scale application tests at a plant with close operation between trains with the KADY process and trains without the KADY process. The Springfield test results can be interpreted as having no net solids reduction, or a 12% solids reduction with KADY. There is no fundamental basis to believe that the KADY operation would cause a higher observed solids yield. Assuming a 12% reduction, a power requirement of 10 Hp/Mgal-d to apply the KADY process, and values given for KADY equipment costs in papers by Springer and Higgens (1999), a rough analysis was done to determine the required costs of sludge processing and hauling (in \$/dry ton) to make the KADY process attractive. The equipment cost was taken as \$33,000 per unit plus 50% for engineering and installation and was amortized at 8% interest over five years. The results are shown in Table 5-3.

Table 5-3. Comparison of Estimated Sludge Reduction Costs Using the KADY Process to Sludge Processing and Hauling Costs to Break Even.

Parameter	Value
Secondary Solids Produced	800 lb/d
Solids Reduction at 12% Efficiency	96 lb/d
Equipment Amortization	\$22/day
Cell Disruption Equipment Power	\$14/day
Additional Aeration Power	\$3/day
Maintenance	\$7/day
Total	\$46/day
Cost in \$/dry ton	\$960

Power cost was assumed to be 8¢/kwh. To determine the increased aeration costs for additional oxygen transfer to oxidize the COD produced from solids destruction, the assumptions are 0.8 g O_2 /g COD, 1.4 g COD VSS destroyed, and 1.5 kg O_2 transfer/kwh. The equipment maintenance cost was taken at 10% of the capital cost per year (Springer et al., 1996).

The sludge process costs would have to be about \$960/ton to make the mechanical disintegration process economically attractive, which is well above typical values of \$100-300/ton for sludge processing and hauling. If the equipment capital cost is ignored, the operating cost is about \$460/ton, which is still excessive compared to conventional costs. Assuming \$300/ton for sludge processing, the KADY process would have to produce a sludge reduction of 69% to break even, including the estimated equipment cost or a 35% reduction without the equipment cost. Assuming a biomass fraction of 75% (Chapter 1.0), the percent biomass reduction would have to be 46–92%. Test results provided would suggest that these levels of biomass reduction are not achievable. Thus, while the mechanical disintegration process appears technically feasible, the energy consumption may be too high to result in an overall economic benefit.

Another way to look at the economics of biomass reduction is to consider the cost of solids processing and disposal and to relate that to an equivalent energy cost per unit of solids destroyed. At a \$100-300/ton cost for solids processing and disposal, the sludge reduction technology must cost less than 5-15¢ per lb of solids destroyed. Assuming 8¢/kwh for energy, the technology must consume less than 0.63-1.9 kwh/lb solids destroyed, if the only costs are for energy.

Energy requirements for 15% solids destruction by various methods have been presented by Müller (2000). His values are converted to kwh per lb of solids destroyed and summarized in Table 5-4. The evaluation suggests that HPH processes may have potential for an economic reduction of biomass solids, but that ultrasonic homogenizers and thermal treatment may not be economically attractive. For the HPH process, results from the literature do not report the energy required to operate the systems. However, laboratory tests results showed possible cell destruction in the range of 25–80% biomass reduction. Larger scale tests are needed to further assess HPH processes.

The energy costs for the thermal process are excessive and suggest that the process is not economically attractive. Other disadvantages for the thermal treatment are the space requirements (24-hour incubation time) and operational and maintenance requirements for a more complex system.

Table 5-4. Estimated Energy Requirements for Mechanical and Thermal Cell Disruption (Data from Müller, 2000).

Cell Disruption Process	Est. Energy Requirements, kwh/lb solids destroyed
High Pressure Homogenizer	0.3
Stirred Ball Mill	0.6
Ultrasonic Homogenizer	4.4
Thermal	2.3

5.6 Research Needs

Mechanical and thermal processes are able to cause cell disruption and thus reduce biomass production. A major issue is the relationship between the energy input and degree of cell disruption for various processes. An additional issue is how the performance and energy needs may change as processes are scaled up from bench-scale studies to full-scale applications. Of the processes evaluated in this category, the HPH process appears to be the most attractive in terms of potential cell destruction at an acceptable energy input. However, research is needed at pilot and full scale to better assess the efficiency using realistic equipment sizes.

CHAPTER 6.0

CELL DISRUPTION PROCESSES BY CHEMICAL METHODS

6.1 Description of Process

Chemical additions have been used to create stress or oxidative conditions to cause cell disruption and release of intracellular components for further degradation. Sludge has been treated with acids, bases, ozone, chlorine, and enzymes, as examples of reduction methods applied. A common approach has been to apply the chemical to a portion of the RAS flow, where the activated sludge solids are at a higher concentration.

Ozonation of RAS has been evaluated by a number of investigators and has been shown to reduce solids production. A schematic of a treatment scheme with ozonation of RAS is shown in Figure 6-1. Other chemicals may also be applied in lieu of ozone for alternative biomass reduction technologies.

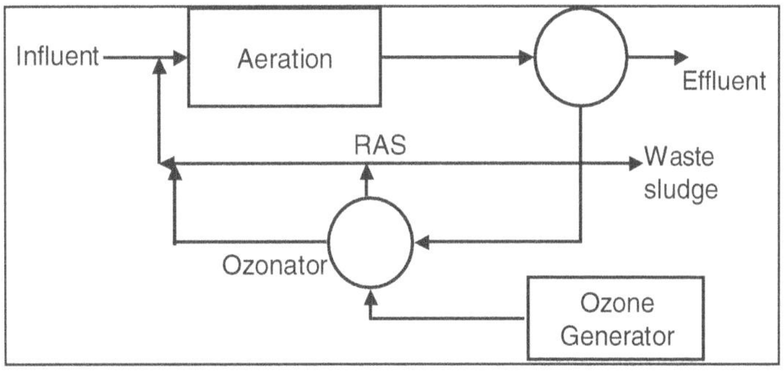

Figure 6-1. Schematic of Cell Disruption Process Using Ozonation.

Kurita Industries in Japan has developed a RAS ozonation process and have shown sludge reduction in bench-, pilot-, and full-scale tests (Yasui et al., 1996; Kamiya and Hirotsuji, 1998; Sakai et al., 1997). A portion of the RAS flow is directed to an ozone contactor with a detention time of approximately 15 minutes. The side stream flow to the contactor is such that the aeration tank solids are typically cycled through the contactor once every three-four days. Another guideline for ozone contacting is that the amount of sludge exposed to ozone is about three times the amount of sludge wasted daily (Sakai et al., 1997). The ozone contactor has a height of about 5 m (1.52'), and ceramic diffusers disperse ozone into the mixed liquor volume. About 90% ozone utilization has been claimed. Exhaust gas from the contactor is treated by a catalyst and activated carbon. An ozone generator is installed for continuous ozone production, with ozone dosing rates in the range of 0.02–0.05 g O_3/g VSS.

The substrates released from mixed liquors exposed to ozone have been shown to be biodegradable and able to promote cryptic growth (Egemen et al., 1999). In parallel bench-scale studies with a control activated sludge reactor versus an activated sludge reactor with ozone treatment, the effluent soluble COD concentrations were similar (50 versus 69 mg/L, respectively) even though the solids in the ozone treated reactor were reduced by 40–60%. Ahn et al. (2002b) showed that ozone treatment of RAS could provide a readily available biodegradable carbon source to drive denitrification rates in an anoxic aerobic system. Specific denitrification rates of 0.01 to 0.08 g NO_3-N/g VSS–d were achieved with ozonated RAS, which compares well to reported specific rates in the range of 0.03 to 0.10 g NO_3-N/g VSS–d for pre-anoxic reactors with influent municipal wastewater.

The idea of using sodium hydroxide for high pH treatment of activated sludge to reduce biomass production was proposed by Gaudy et al. (1971) many years ago in their work with extreme SRT systems. Penaud et al. (1999) showed that with NaOH additions of 3.5–13.9 g/L with pH values of 10–12, the VSS reduction of an industrial wastewater mixed liquor could be in the range of 20–40%, respectively. High pH has been shown to disrupt the cytoplasmic membrane of bacteria cells to release intracellular material (Mendonca et al., 1994). This technology, though effective, was not pursued further due to the lack of any known applications and the negative impacts of greatly increasing the total dissolved solids content of treated wastewater. Low pH treatment has also been proposed and found effective for solids destruction at high acid doses.

Enzyme additions or "additives" by various vendors have led to many claims for improved wastewater treatment and less sludge production. Because of conflicting opinions about the success of this alternative, the wide range of products, and the ability for engineers or wastewater treatment plant operators to carry out their own controlled test, this technology for sludge reduction was not pursued in this work. There was little expectation that a general finding could be developed from this avenue. However, an illustrative example that enzyme additives do have potential for sludge reduction is found in the more scientifically based work by Parmar et al. (2001). They showed that a mixture of cellulase, protease, and lipase enzymes could accelerate solids destruction to reduce sludge production. They did note that enzymes are relatively expensive and further improvements in enzyme production are needed to make the technology more attractive.

Chlorination also has been used in activated sludge processes in bench-scale studies with the objective of reducing sludge production. Similar to the ozone treatment process, sludge from the bioreactor was treated through a chlorine contact chamber. Control reactors with no chlorine addition also were operated in parallel (Saby et al., 2002; Nolasco et al., 2002). At a chlorine dose of 0.07 g Cl_2/g MLSS, a 65% reduction in solids production was observed. Cell counts were provided to confirm the loss of biomass. The investigators did raise concerns about trihalomethane formation, the need to monitor and possibly treat gas emissions, and to further investigate chlorine compounds in the effluent liquid. Because of these potential toxic production issues, the current reluctance to use chlorine at wastewater facilities, and lack of application of this method, this technology was not evaluated further in this study.

6.2 Process Fundamentals and Biomass Reduction Mechanism

Disinfection mechanisms described by Crites and Tchobanoglous (1998) explain how oxidants such as ozone or chlorine can disrupt bacteria cells and release intracellular material. They note that strong oxidants damage the cell wall and that such damage results in cell lysis and death. In the ozonation applications the ozone dose is applied at a level sufficient to disrupt the cell, but not at a high enough dose to accomplish a significant amount of oxidation of cell organic material. Additional oxygen was needed in the activated sludge aeration tanks to oxidize released organic substances.

6.3 Reported Process Performance

A number of studies clearly show that ozone treatment of RAS reduces solids production. Table 6.1 summarizes the results from studies that involved a parallel control activated sludge system without ozone treatment. The general operating guideline used in these studies to fix the RAS flow rate to the ozone contactor was to process daily a mass of mixed liquor equal to three times the mass wasted from the control system without ozone treatment.

The solids reduction efficiency is shown to vary with the ozone dose in the ozonator in terms of g O_3 applied/g TSS fed. At a dose of 0.05 g O_3/g TSS applied, a more common operating condition (Yasai et al., 1996), 70–80% solids destruction was achieved. At this level the amount of ozone used per g TSS destroyed was in the range of 0.20–0.40 g O_3/g TSS destroyed. The results appear reasonable based on the care taken to operate parallel control reactors and the long-term testing periods used (weeks to months).

One unexpected and unexplained result is the 99% solids reduction for the full-scale municipal wastewater study with raw sewage at a 0.04 g O_3/g TSS dose reported by Sakai et al. (1997). The high level of solids destruction does not fit with fundamental principles. The fraction of biomass for activated sludge treatment of raw wastewater should be no more than 70–80%. Thus, 99% destruction is not possible for the cell disruption mechanism expected.

In another pilot plant study with activated sludge treatment of municipal wastewater reported by Ahn et al. (2002a), the results were characteristic of those summarized in Table 6.1. Increasing solids destruction occurred at higher ozone doses and 70% solids reduction was consistently achieved. However, the ozone dose reported was shown as a factor of ten higher as g O_3/g TSS. Whether this is a typographical or calculation error, cannot be determined from the results provided.

Tests by Kamiya and Hirotsuji (1998) and Huysmans et al. (2001) showed that intermittent ozone treatment of the RAS at a higher dose provided greater solids destruction for the same amount of ozone used than a constant ozone application at a lower dose.

Table 6-1. Summary of Sludge Reduction Performance with Ozone Treatment.

Test Condition	Wastewater	Flow rate	A.S. SRT, d	Contactor ozone dose, g O_3/g TSS	g O_3/g TSS destroyed	Percent solids Reduction*	Reference
Bench-scale	Synthetic (Pentone)	8 L/d	N/A	0.05 0.10 0.20	0.21 0.36 0.65	73 83 90	Yasai and Shibuta, 1994
Full-scale	Municipal Raw screened	430m^3/d	720	0.02 0.04 0.04	0.15 0.17 0.21	36 77 99	Sakai et al., 1997
Bench-scale	Municipal Primary effluent	15 L/d	10.0	0.05	—	70	Deleris et al., 2002
Bench-scale	Synthetic	10 m^3/d		0.02	0.14	50	Huysmans et al., 2001

In all of the tests no significant deterioration in effluent quality was found. Yasui et al. (1996) reported a slightly higher effluent TOC concentration at less than 2% of the solids destroyed and showed effluent TSS concentrations below 10 mg/L for tests with numerous industrial wastewaters (oil refinery, food industry, petrochemical, and pharmaceutical) and with domestic wastewaters. No negative effect on nitrification performance or maintenance of an autotrophic nitrifying bacteria population was shown by Deleris et al. (2002), Huysmans et al. (2001), Sakai et al. (1997), and Ahn et al. (2002a). At first glance this seems surprising because ozonation destroys bacterial cells, including autotrophs. However, the SRT of the ozone treated system was much greater than the control system due to the solids reduction and operation with the same aeration HRTs and elevated MLSS concentrations, so that a large nitrifying bacteria inventory was possible.

Many of the studies reported improved sludge settling properties with ozone treatment. Deleris et al. (2002) developed a high SVI (300 mL/g) in their completely mixed activated sludge reactor before ozone treatment. With ozone treatment the SVI decreased to 180 mL/g over a 30-day period. Microscopic observations showed a dramatic decrease in the presence of filamentous organisms, which is likely due to the application of the ozone to the RAS. A strong oxidant, such as chlorine, has often been applied to RAS for filamentous bulking control (Tchobanoglous et al., 2003). In the pilot plant study by Ahn et al. (2002a) the mixed liquor SVI was improved from 100 to 80 mL/g with ozone treatment. Kamiya and Hirotsuji (1998) showed that the SVI could be maintained below 100 mL/g with ozone treatment, but exceeded 400 mL/g without ozone treatment, suggesting the proliferation of filamentous organisms.

6.4 Plant Retrofit Considerations

The ozonation process can be applied for plant retrofit applications in a reasonably straightforward fashion. A side stream from each RAS line would need to be pumped to an ozone contactor and an ozone generator would have to be provided on site. Sufficient space would be needed for the ozone contactor, but with its relatively short detention time requirement of about 15 minutes, space may be available at most sites.

6.5 Assessment of Process

Ozone treatment of RAS for solids reduction has been clearly shown to be an effective biomass reduction technology and has been shown in pilot plant and full-scale tests to be a feasible process. A unique feature of the process is its operational flexibility. The degree of solids destruction can be controlled by the ozone dose (in g O_3/g TSS) applied to the ozone contactor, the fraction of solids inventory fed to the contactor, and the use of the intermittent versus continuous treatment.

Additional operator skill would be needed to operate and maintain the ozone generator and ozone contactor, but this appears to be within normal capabilities based on previous applications using chemical oxidants such as chlorine or ozone for disinfection at wastewater treatment facilities. The improved settling characteristics of the activated sludge mixed liquor would provide a positive impact on plant operations by preventing bulking sludge and providing a more stable operation.

A major issue for the application of ozonation for biomass reduction is costs. Ozonation was rated highly by the project technical team and project steering committee and a cost evaluation was recommended and carried out. The cost analysis is provided in Appendix B and is summarized here (Table 6-2) for a 5 Mgal/day wastewater flow for a municipal and industrial wastewater facility. The municipal wastewater system design was based on primary and secondary activated sludge treatment of a typical domestic wastewater with an influent BOD and TSS concentration of 200 mg/L. The industrial wastewater treatment scenario consisted of secondary activated sludge treatment only for a stronger highly soluble wastewater with an assumed BOD of 600 mg/L. Thus more biomass production occurred in the industrial wastewater process, which provided a greater opportunity for more biomass production. For both applications 60% biomass reduction was assumed for ozonation of the RAS, and the ozonation design was based on values reported in previous tests for ozone dose, frequency of RAS exposure, and ozone contact time.

The cost analyses also considered two options for sludge processing and disposal or reuse. Option A was for a conventional approach of thickening, digestion, dewatering, and biosolids reuse. Option B assumed solids thickening prior to disposal in a sanitary landfill. The economic analysis showed that the cost of biomass reduction by ozonation greatly exceeded the cost of processing the waste sludge by the plant conventional methods. The hauling distances for ultimate solids disposal had to be in the range of 270–830 miles to justify the cost of using ozone for sludge reduction (Table 6-2).

Table 6-2. Summary of Cost Evaluation of Using RAS Ozonation Process for Biomass Reduction Described in Appendix B.

Type of Plant	Sludge Handling Option	Ozone Costs, $/Ton Dry Solids	Plant Processing Costs, $/Ton Dry Solids	Hauling Distance Needed to Make the Ozone Process Break Even, miles
Industrial	Option A	746	366	271
Industrial	Option B	746	135	436
Municipal	Option A	782	315	646
Municipal	Option B	782	143	828

Other investigators found a favorable economic condition for biomass reduction with ozone. In these cases the ozonation sludge reduction technology was compared to sludge incineration as the ultimate residuals handling process. All the investigators admitted that the cost analyses were based on simple estimates. Table 6-3, from Huysmans et al. (2001), shows that the RAS ozonation process is competitive with conventional treatment with sludge incineration with ozone doses sufficient to cause zero sludge growth or a 50% reduction in sludge production. Further full-scale testing and a more detailed analysis would be needed to compare the process more closely with sludge incineration.

Table 6-3. Summary of Cost Analysis by Huysmans et al. (2001) Comparing Sludge Handling Costs for RAS Ozone Treatment and Conventional Treatment with Incineration. *

Process Effects	Ozonation Costs, Euro Dollars		Conventional Costs, Euro Dollars
	Zero Sludge	50% Reduction	
Ozone Treatment	430	210	—
Additional Aeration	50	25	—
Dewatering		90	180
Transport		25	50
Residual Incineration		125	250
Total	480	480	480

* The cost of ozone based on equipment installation and depreciation and energy cost was estimated at 3.75 Euro Dollars per kg O_3.

In contrast to the cost analysis by Huysmans et al. (2001), Ahn et al. (2002a), and Yasui et al. (1996) produced numbers showing that the RAS ozonation process reduced the cost of sludge processing and disposal by 50% compared to incineration for application in Korea and Japan, respectively. These analyses suggest that for cases where ultimate disposal or reuse of processed sludge is difficult so that incineration is a key alternative, the RAS ozonation process should be evaluated.

One disadvantage for significant sludge reduction and minimal sludge wasting from the activated sludge process is that it prevents using EBPR processes to meet effluent permit requirements of low phosphorus concentrations. With EBPR processes, the phosphorus is removed from the system via the wasted biological sludge. By destroying biomass, less phosphorus can be removed by this method, so that phosphorus removal would have to rely on chemical precipitation.

6.6 Research Needs

A significant number of bench, pilot, and full-scale tests have been done that have proven the feasibility of the ozonation process for sludge reduction, developed design and operational guidelines, and assessed the effect on treatment performance. However, two areas of further research have been identified. The first is to evaluate the effect of ozonation on the nonbiodegradable volatile suspended solids (nbVSS) in the system. For some test results presented, the level of solids destruction suggests some destruction of nbVSS. Because nbVSS can accumulate in the system at significant levels (30–50% of MLVSS) and produce a net sludge production, this is an important parameter to evaluate.

The second issue is to investigate the optimization of the system SRT with RAS ozonation. Tests should be done with parallel operation of a system with ozone treatment and conventional treatment with equal operating SRTs. In previous work the parallel systems were operated with equal aeration times and it appeared that the ozone treated systems were operated with a longer SRT. An advantage not claimed by the previous investigators is the fact that for a given SRT the reduced sludge reduction will reduce the aeration volume requirements or allow treatment of a greater flow rate for the same SRT. This could provide a significant cost advantage for the ozone treatment process not previously recognized. Presumably the SRT that can be used with ozone treatment may have to be greater than that without ozone treatment to offset the effect of bacteria kill, especially for nitrification systems. However, research is recommended to evaluate how much greater the SRT should be to provide a stable process and to allow for full nitrification.

CHAPTER 7.0

Chemical Uncoupling Processes

7.1 Description of Chemical Uncoupling Processes and Fundamental Mechanisms

Research on methods for stimulating uncoupling of carbon oxidation from cellular metabolism is a recent development in wastewater treatment. This concept has the appeal that biomass production is avoided from the outset, compared to methods such as predator enhancement that rely on later processes to mineralize biomass that is produced by oxidation of organic matter. Two approaches have been taken toward stimulating such uncoupling: chemical uncoupling and energy spilling under stressful environmental conditions.

Chemical uncoupling occurs when membrane-active chemicals such as protonophores cause leakage of the proton gradient across the cell membrane. The proton gradient is responsible for ATP generation. Under these conditions oxidation of organic carbon continues, but ATP generation is inhibited and cell growth is slowed.

Another approach is to take advantage of naturally inefficient microbial growth patterns under stressful environmental conditions, such as high-temperatures and growth limitation by nutrients other than carbon. The second approach is generally limited to industrial wastewaters of unusual compositions or high temperature (Lapara and Alleman, 1999; Lapara et al., 2000). If the uneconomical heating of wastewater is to be avoided, autothermal operation is required, in which the energy released from metabolism is used to heat the reactors. Analysis indicates that autothermal operation requires a COD removal in excess of 20,000 mg/L (Lapara and Alleman, 1999). This requirement effectively rules out the application of thermophilic approaches to sludge reduction to municipal and most industrial wastewaters.

Alternatively, heterotrophic growth can be reduced if nutrients other than the carbon and energy source are limiting. So-called "energy spilling" occurs under carbon-sufficient conditions (Liu 1998). Since the removal of organic carbon is the primary goal of most biological wastewater treatment processes, microbial growth is limited by carbon in the final reactor stage. Thus, nutrient-limited energy spilling is not practical except in the presence of very high carbon levels compared to nutrient availability.

Nutrient limitation has been applied to pulp mill effluents by restricting the amounts of nitrogen and phosphorus that are added, but problems with poor biomass settleability have been encountered due to excessive formation of polysaccharide slime. Recent experiments with a moving bed biofilm process combined with activated sludge operated under nutrient limiting conditions produced a yield of 0.09 g TSS/g COD removed, combined with 80-90% COD removal and good sludge settleability (Welander et al., 2002). Slimy biofilm produced in the first stage was degraded in the activated sludge (AS) stage. This process has the advantage of avoiding nutrient release from biomass mineralization. Costs associated with nutrient addition are also avoided.

In recent years a handful of laboratory studies have been conducted on the effects of chemical uncouplers on activated sludge. About 20 uncoupling chemicals have been investigated (Table 7-1). All of the studies found that the addition of selected chemical uncouplers in sufficient quantity can result in a significant yield reduction. The pertinent parameters of these studies are summarized in Table 7-2.

Table 7-1. Uncoupling Chemicals Tested for Their Effects on Activated Sludge.

Uncoupling Chemical	Reference
2,4-dinitrophenol*	Rich and Yates, 1955
2,4,5-trichlorophenol*; *o*-nitro-*p*-chlorophenol*; 2,4,6-tribromophenol*; 2,6-dibromo-4-nitrophenol*; and 2,4-dichlorophenol*	Okey and Stensel, 1993
Vancomycin; quinacrine; Congo red; Chlorpromazine; erythromycin*; rotenone*; trypan blue*; 2,4-dinitrophenol*.	Mayhew and Stephenson, 1998
2,4,5-Trichlorophenol*; 4-chloro-2-nitrophenol (CNP)*; carbonyl cyanide m-chlorophenylhydrazone*; 3,3',4',5-tetrachlorosalicylanilide*; 2,4-dichlorophenol*; carbonyl cyanide p-(trifluoromethoxy) phenylhydrazone*; phenylglyoxal*; pyridoxal-5-phosphate*; quercetin*; 2,4-dinitrophenol (DNP); gramicidin D; 8-anilino-naphthalene sulfonic acid (ANS)	Strand et al., 1999
Para-nitrophenol*	Low et al., 2000
Zinc*; 2,4-dinitrophenol*; 3,3',4',5-tetrachlorosalicylanilide*	Liu, 2000; Chen et al., 2002

*Positive uncoupling results observed.

7.2 Reported Process Performance

Uncoupling chemicals have been tested only at bench-scale, and most uncoupling chemical effects have been conducted in short-term batch tests. Three chemicals have been tested in prolonged, continuous-flow, bench-scale activated sludge bioreactors: para-dinitrophenol (Pnp); 2,4,5-trichlorophenol (TCP); and 3,3',4',5-tetrachlorosalicylanilide (TCS) (Strand et al., 1999; Low et al., 2000; Chen et al., 2002).

Depending upon the chemical used and the concentration applied, the initial effects on observed yields can be dramatic. As shown in Table 7-2, applications of TCP and TCS can reduce sludge yields by 50–60% at concentrations of 8.5 mg/L TCP and 1.0 mg/L TCS.

7.3 Plant Retrofit Considerations

Retrofit requirements for use of chemical uncouplers would be modest. Chemical storage and dosing equipment would be needed. Additional unit processes may be required in order to reduce the effluent concentration of these potentially toxic chemicals to acceptable levels. The most significant requirement would be the need for additional aeration. The specific OUR increased proportionately to the reduction in sludge yield (Strand et al., 1999; Chen et al., 2002), thus additional aeration capacity may be required to prevent the development of anaerobic conditions. Monitoring of process and effluent levels of the uncoupling chemical would place additional analytical burdens on plant operations. Generally these chemicals are analyzed by gas chromatography. Both of the chemicals (TCS and TCP) shown to have the greatest effects at low concentrations are chlorinated and may be best analyzed using electron capture detection.

7.4 Assessment of Process

Two issues stand out in the analysis of uncoupling chemicals for sludge reduction: the long-term sustainability of the effects of uncouplers on an activated sludge population, and the acceptability of the potentially toxic uncoupler chemical in the wastewater effluent.

The durations of continuous flow bench-scale reactor tests to date have ranged from 20–85 days. During the longest of these tests, the activated sludge culture became acclimated to the uncoupling chemical (TCP), and the uncoupling effect on yield was reduced (Strand et al., 1999). None of the other tests were conducted long enough to clearly demonstrate that acclimation can be avoided in completely mixed reactors.

Surprisingly, only one of the studies to date has addressed the important question of the level of the uncoupler chemical in the effluent. In this test the concentration of the uncoupler, TCP, was directly correlated with the reduction in sludge yield; i.e., as the TCP concentration in the effluent decreased, the sludge reduction also decreased (Strand et al. 1999). Although Liu (2003) noted that TCS has been used in personal care products in the past, it is unlikely that regulatory agencies will accept discharge of chlorinated aromatic compounds such as TCP and TCS in the ppm range shown to have the desired sludge reduction effect.

While the concentrations of TCP and TCS required in the laboratory experiments to date suggest that use of these uncoupling chemicals may be an economical method for sludge reduction, bench-scale studies and the uncertainties surrounding activated sludge acclimation preclude a meaningful cost analysis.

7.5 Research Needs

Future research should focus on methods to avoid acclimation of the activated sludge cultures to the uncoupling chemical. Such methods might include the use of plug flow bioreactors and alternating types of uncoupling chemicals. A long-term test of TCS is needed to determine whether acclimation can be avoided for this chemical. All of the tests of uncoupling chemicals to date have used synthetic wastewater. The effects of uncouplers on the treatment of actual municipal wastewater need to be determined. Finally, all future studies of chemical uncouplers should include analysis of effluent concentrations of the chemicals.

Chemical uncoupling sludge reduction technology has some potential as an economical and general approach to sludge reduction, but further laboratory study is required at this time.

Table 7-2. Chemical Uncoupling Studies.

Scale	Duration	Flow Rate, L/d	Chemicals Used	Average Net Solids Yield	Effluent Quality Parameters	Comments	Reference
Bench-scale, 1.0 L bioreactor with recycle, synthetic wastewater, HRT = 3.5 h, SRT = 22-29 d	26 d	7	100 mg/L p-nitrophenol	0.24-0.3 mg/mg	TOC 90 mg/L before, 270 mg/L after treatment. No measurements of uncoupler in the effluent	Disrupted settling characteristics. Protozoa disappeared from system. Filamentous bacteria proliferated. Diversity of bacteria decreased as shown by changes in Rrna.	Low et al., 2000
Batch tests and bench-scale activated sludge with 3 L porous pots. HRT = 5.5 h. SRT NA	Up to 55 d	0.55	2,4-dinitrophenol, 35 mg/L. Batch tests of the effects of 8 compounds on activated sludge respiration.	0.30±0.05 mg SS/mg BOD compared to 0.42±0.09 for parallel control	7.3 mg/L BOD compared to 5.3 for control, no difference. No measurements of 24DNP and effluent. BOD removal = 95%	No discussion of chemical effects on settling qualities.	Mayhew and Stephenson, 1998
Bench-scale, short-term batch tests of the effects of on AS respiration	NA	NA	zinc and 2, 4-dinitrophenol	0.25 mg MLSS/mg glucose at 0.16 mg 24DNP/mg MLSS compared to 0.6 mg MLSS/mg glucose. For zinc, yield fell from 0.4 to 0.19 mg MLSS/mg glucose.	NA No measurement of uncoupler concentration.	Artificial wastewater. 1 mg glucose/mg MLSS. Biomass concentration = 0.05-0.2 g/L. No discussion of chemical effects on settling qualities.	Liu, 2000
Bench-scale and Batch tests of 12 chemicals. 2 L CMAS system. 5 d SRT, 3.5 h HRT, SS = 0.05-0.2 g/L.	85 days for CMAS	13.1	2,4,5-trichlorophenol in CMAS. Batch tests of 12 chemicals. Best uncouplers were 2,4,5-trichlorophenol and 4-chloro-2-nitrophenol	0.22 mg MLSS/mg COD at 8.5 mg TCP in effluent compared to 0.50 mg MLSS/mg COD for the parallel control.	2-50 mg SCOD/L for both treated and control. TCP in effluent ranged from <1 to 8 mg TCP/L, with lower values at end of test period.	Specific oxygen uptake rate was inversely related to the yield.	Strand et al., 1999

Table 7-2. Chemical Uncoupling Studies (continued).

Scale	Duration	Flow Rate, L/d	Chemicals Used	Average Net Solids Yield	Effluent Quality Parameters	Comments	Reference
Batch & Bench-scale AS system. 25 L SBR, HRT = 8 h, SRT = 7 d.	3 mo.	NA	3,3',4',5-tetrachlorosalicylanilide (TCS)	Observed yield fell from 0.45 with no TCS to 0.12 mg biomass COD/mg soluble COD at 0.8 mg/L TCS	NA. TCS concentrations not measured	Cell ATP declined more rapidly in the presence of TCS. Cell carbohydrate storage was unaffected by TCS.	Chen et al., 2000
Batch and Bench scale AS system	30 d	NA	3,3',4',5-tetrachlorosalicylanilide (TCS)	Observed yield = 0.57 mg SS/mg COD control, 0.42 at 0.5 mg/L TCS and 0.30 at 1.0 mg/L TCS	Not reported directly. Substrate removal unaffected. No measurement of TCS in effluent	Total & respiring bacteria enumerated by DAPI & CTC staining. Respiring cells increased from 3.9% in the control to 7.6% in the culture with 1 mg/L TCS.	Chen et al., 2002

CHAPTER 8.0

PREDATOR ENHANCEMENT PROCESSES

8.1 Description of Processes

The role of predators in controlling suspended solids in the effluents of trickling filters has been appreciated for several decades. In the last decade several groups have studied the effects of predator enhancement on carbon mineralization and sludge production. Two concepts have emerged that have potential for practical predation enhancement: addition of protozoa after growth in separate bioreactors (Rensink and Rulkens, 1997; Cox and Deshusses, 1999), and separation of the bacterial oxidation stage from a subsequent predation stage (Lee and Welander, 1996a, 1996b; Welander et al., 2000). While the former approach is unlikely to be economical due to its requirement for the offline cultivation of predators, the latter method has been demonstrated at full-scale and is more attractive economically. This technology, pioneered by Welander and others at Lund University and ANOX AB, is based upon the concept of a bacterial bioreactor operated without recycle and at an HRT just below the washout rate for the bacterial culture (Figure 8-1). Carbon oxidation occurs primarily in this reactor. The bacteria in the effluent from the oxidation reactor are consumed in a second reactor with a high enough SRT to sustain a protozoan/metazoan population. The second reactor may be a biofilm reactor (Lee and Welander, 1996b), activated sludge system with sludge recycle (Welander et al., 2000), or a membrane bioreactor (Ghyoot and Verstraete, 1999).

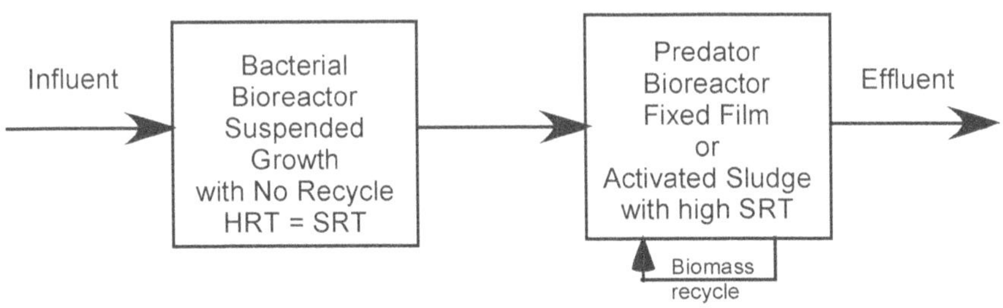

Figure 8-1. The Low Sludge Production (LSP) Process with Two-Stage Predation Enhancement.

The two-stage bacterial/protozoan bioreactor system has shown promise in laboratory tests with artificial high strength wastewaters and with industrial wastewaters (Table 8-1). Questions remain regarding its applicability to municipal wastewaters with their low soluble

COD, low temperatures, and highly variable bacterial input. We are aware of only one test with a municipal wastewater, a study at bench and pilot scales conducted by Christensson and Welander (2001) at ANOX AB. In this study a municipal wastewater was found to be slowly degraded in the first stage, requiring a retention time of 4 h to reach an average soluble COD (SCOD) removal of 50% in the laboratory. In the pilot-scale study, COD removal was only 30% at an HRT of 1.5–2.4 h. Bacterial growth in the first stage was limited. Importantly, there was no reduction in SS in the pilot-scale second stage (AS with recycle). Effluent SS with a low sludge production process (LSP) was unchanged from the operation as conventional AS. While overall COD removal was good with LSP, the yield did not decrease significantly compared to the conventional system, 0.30 g TSS/g COD versus 0.35 g TSS/g COD. Thus, application of predator enhancement technologies such as LSP does not appear to be applicable to low strength, low temperature wastewaters such as municipal sewage.

8.2 Process Fundamentals and Biomass Reduction Mechanism

The predator enhancement process relies upon biological reprocessing of the bacterial biomass produced during consumption of wastewater organic matter in a second stage bioreactor designed to maintain a high biomass of protozoa or metazoa. This process differs fundamentally from anaerobic processes and chemical uncoupler processes because in the former biomass that is efficiently produced from soluble organic matter must be subsequently mineralized, while in the latter, biomass production is avoided in metabolism of the organic matter.

8.3 Reported Process Performance

As shown in Table 8-1 yield reductions with LSP are substantial, rangin from 20–90%, with an average of about 50%. These sludge reductions are possible only for moderate to high strength wastewaters.

8.4 Plant Retrofit Considerations

A bacterial bioreactor with a low HRT (less than 6 h) is needed. Existing selector tanks may suffice. This reactor requires no recycle and should have sufficient aeration to oxidize most of the influent COD. The HRT of the bacterial bioreactor must be adjusted to promote dispersed growth and to avoid filamentous or flocculent growth.

The predator stage may be a conventional activated sludge with recycle, a biofilm reactor such as trickling filter, or a membrane bioreactor. SRTs in the predator stage have been high, ranging from 10–122 days.

Due to the increased biomass mineralization, aeration requirements are higher and additional aeration equipment may be needed. Operational considerations include increased demands on operator skills, especially in microscopic examination of the bacterial dispersion and on predator counts.

8.5 Assessment of Process

LSP has the greatest potential of existing aerobic processes for producing sludge reductions equivalent to anaerobic processes for moderate to high strength wastewaters. Demonstration at full-scale provides this assessment with substantial support. Costs have not been explicitly evaluated in publication, but appear to be lower with savings in sludge handling offsetting higher aeration costs.

One problem with LSP technology—nutrient release—is common to technologies that rely upon mineralization of biomass in the wastewater treatment process stream. Oxidation of bacterial biomass carbon may produce high levels of soluble nitrogen and phosphorus in the effluent, possibly exceeding regulatory levels (Stuart et al., 2001). Welander et al. (2000) prevented significant N and P release by limiting the addition of nutrients to the wastewater from the Norske Skog Folla CTMP mill in Norway, but laboratory experiments have shown N and P release to be a potentially serious problem for this technology (Ghyoot and Verstraete, 1999).

The need to avoid filamentous and flocculent growth in the bacterial reactor may limit applications.

8.6 Research Needs

More pilot and full-scale applications with actual wastewaters are needed to assess the types of wastewaters and operational conditions that favor dispersed growth in the bacterial reactor and how nutrient release can be minimized.

Process modeling is needed to allow rational design of the bacterial stage HRT and predator stage SRT. While the process is apparently not applicable to municipal wastewaters in its present form, further experimentation could define the minimum wastewater strengths and temperatures to which predator enhancement can be applied.

Table 8-1. Predator Enhancement Technologies Using Two Stage Treatment Processes.

Scale	Dura-tion, d	Flow, L/d	Test set up	Controls	Wastewater characteristics	Effluent quality parameters	Solids yield	Comments	Reference
Bench-scale 0.1-0.8 L bacterial stage, 0.5-0.8 L predator stage. Bacterial HRT=1-12 h. Predator HRT=6-24 h	15-125	0.8-**!Missing Operator**2.4	3 predator bioreactors—suspended carrier biofilm, stationary support biofilm, & activated sludge.	Bacterial stage, extended aeration & high rate systems (0.20-0.30 g TSS/g soluble COD removed	5 industrial wastewaters, bleached kraft mill, unbleached sulfate mill, recycled fiber mill, & CTMP. 450-6620 mg SCOD/L. 500-1820 mg TCOD/L	57-91% removal of COD. Effluent P & NO_3-N high	0.01-0.46 g effluent TSS/g COD	Bacterial HRT selection critical. Not demonstrated for municipal sewage.	Lee and Welander, 1996a
Bench-scale 0.8 L. Bacterial stage HRT = 10 h	12 or 160	1.9	2 systems 1st stage = CM reactor, 2nd stage = biofilm reactor 50% filled with PE particles.	Bacterial stage only. 0.51 g/g acetate 0.29 g/g MeOH	Synthetic wastewater based on acetic acid or methanol as sole sources of carbon 1000 mg COD/L	92-94% COD removal effluent P & NO3-N high.	0.17 g TSS/g COD acetate. 0.048 g TSS/g COD for MeOH	Large floc and heavy wall growth with acetate, inhibited grazing. Protozoan growth with methanol. Stable predator reactor dominated by rotifers > 50 d.	Lee and Welander, 1996b

Table 8-1. Predator Enhancement Technologies Using Two Stage Treatment Processes (continued).

Scale	Dura-tion, d	Flow, L/d	Test set up	Controls	Wastewater characteristics	Effluent quality parameters	Solids yield	Comments	Reference
Bench-scale, 24L, HRT = 0.16 h, Gas flow rate = 15000 m3/h, No recycle	56	5.8	Trickling filters, added protozoa, concurrent liquid & air flows.	Parallel without added protozoa	Mineral medium with 0-3.5 g toluene/m^3 in gas flow stream	Toluene degradation higher for protozoan enriched filter compared to control	0.43 g dry biomass/g toluene, 0.65 control.	Inoculated with *Pseudomonas corrugata* but open cultured. Protozoa were grown separately.	Cox and Deshusses, 1999
Benchscale. 2-2.6 L bioreactors. 1st stage HRT = SRT = 8-34 h. 2nd stage HRT = 17-74 h. SRT = 10-122 d.	11-31	1.4-6.2	Aerated CM reactor HRT=SRT with either activated sludge (CAS) or membrane bioreactor (MBR).	Comparison to first stage alone	Synthetic wastewater, skimmed milk powder. 360, 560, 760, and 1030 mg/L COD 1.4-6.2 L/d	23-165 mg/L SCOD. Increased mineralization in the 2nd stage caused release of N and P	CAS = 0.07-0.29 g SS/g COD. MBR = 0.02-0.3 versus 0.26-0.65 for first stage alone.	20-30% lower sludge production in MBR vs. CAS. Washout of SS in CAS. Nitrifiers decreased & SCOD increased in MBR	Ghyoot and Verstraete, 1999

Table 8-1. Predator Enhancement Technologies Using Two Stage Treatment Processes (continued).

Scale	Duration, d	Flow, L/d	Test set up	Controls	Wastewater characteristics	Effluent quality parameters	Solids yield	Comments	Reference
Pilot-scale, 360 L Activated sludge pilot plants. F/M = 0.1 g COD/g MLSS/d. HRT = 14 h.	11-24	204-408	AS systems with stationary biofilm carrier, amended with oligochaete worm, Tubificiae	Parallel unamended, metazoan suppression by ammonia	Settled domestic wastewater	NA	AS: 0.15 g MLSS/g COD compared to 0.4 control		Rensink and Rulkens, 1997
Full-scale 1300 m^3 bacterial reactor (HRT=SRT= 5-6h) & 15700 m^3 aeration basin.	2 yr +	6000 m^3/d	Conventional AS plant converted to LSP	Yield after conversion compared to yield in conventional AS prior to conversion	CTMP pulp mill effluent (bleached and unbleached). Ave influent COD = 5200 mg COD/L.	82% removal of COD, effluent = ca. 1300 mg COD/L.	0.02-0.05 g TSS/g COD for LSP vs. 0.14-0.22 conventional AS.	Need small, dispersed floc. Filamentous bacteria & high effluent SS problem for high F/M. High N & P in effluent if bacterial release exceeds predator uptake	Welander et al., 2000

CHAPTER 9.0

Extreme Solids Retention Time Processes

9.1 Description of Processes

The extreme SRT process is an activated sludge system that uses special liquid-solids separation methods to maintain extraordinary SRTs (above 50 days). The first evaluation of extreme SRT activated sludge treatment was reported by Gaudy et al. (1971), in which they maintained a laboratory activated sludge process for over two years with minimal solids wasting. Centrifugation of the activated sludge effluent was done manually each day to capture and return the solids to the aeration tank. Though their liquid-solids separation method served their laboratory experimental needs, it was not practical for full-scale activated sludge processes.

With the development and application of microfiltration membrane separation technology for wastewater treatment in the 1990s, a feasible method is now available for almost 100% liquid-solids separation in activated sludge wastewater treatment. Choo and Stensel (2000) reported on solids reduction using an extreme SRT operation with membrane separation in a laboratory reactor. For such systems a high MLSS concentration is needed to minimize the aeration volume requirements. Equation 1.1-1 showed that as the SRT is increased, the aeration volume or MLSS concentration must increase. For membrane bioreactor systems, MLSS concentrations in the range of 8,000–15,000 mg/L have been demonstrated.

9.2 Process Fundamentals and Biomass Reduction Mechanisms

The mechanism claimed for biomass reduction in the extreme SRT process by Gaudy et al. (1971) was the release and metabolism of cellular material as a result of endogenous decay. They assumed that following cell lysis and release of cell material that essentially all of the cellular components could be metabolized by other active bacteria in the system. However, they were not aware that some portion of the lysed cell is not biodegraded by aerobic organisms. This portion is accounted for in current activated sludge models (Barker and Dold, 1995) and consists of nondegradable cell wall material (also referred to as cell debris). About 10–15% of cellular dry weight is expected to remain as cell debris following endogenous respiration.

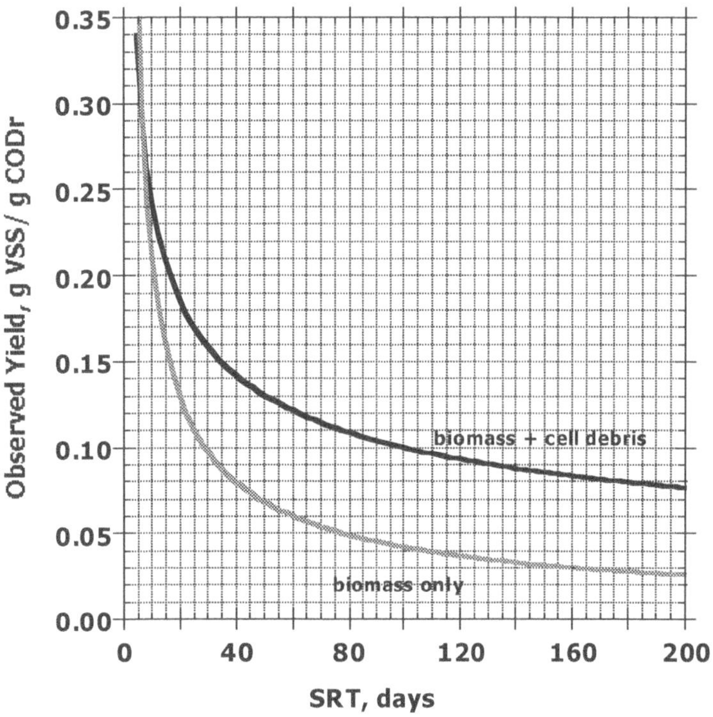

Figure 9-1. Example of Effect of SRT On Observed Biomass Yield and Observed Yield Including Biomass and Cell Debris for Wastewater With No Influent nbVSS.

Equation 1.1-1 can be used to evaluate the net biomass production as a result of extreme SRT processes. The net observed yield, as g VSS/g COD_r for long SRTs is shown in Figure 9-1 with and without the term B in Equation 1.1-1 that accounts for cell debris. For this example the values for the synthesis yield (Y), endogenous decay (k_d), and debris (f_d) coefficients are 0.40 g VSS/g COD_r, 0.08 g VSS /g VSS-d, and 0.15 g VSS / g VSS, respectively. Comparing the observed yields at a five-day and 100-day SRT shows observed yields of 0.30 and 0.10 g VSS/g COD_r, respectively with cell debris included. Without accounting for cell debris, the observed yields calculated are 0.29 and 0.04 g VSS/g COD_r, respectively. With cell debris included the biomass reduction gained by increasing the SRT from five to 100 days is 67% versus 84% if cell debris is ignored. In either case, the example shows that an extreme SRT process can result in a significant biomass reduction.

9.3 Reported Process Performance

Gaudy et al. (1971) reported on a long-term operation of a laboratory reactor treating a synthetic wastewater with 500 mg/L glucose as the carbon source. The reactor had a 16-hour aeration time and settling chamber with an 8-hour detention time. Effluent solids from the settling chamber were captured by centrifugation and returned to the aeration tank. The equivalent COD loading for the aeration reactor was about 0.80 kg COD /m^3-d (50 lb/1000 ft^3-d) and the MLSS concentration ranged from 12,000–16,000 mg/L. About 0.2% of the reactor contents were reportedly used per day for sampling, which would result in an SRT of 500 days. The solids production was reported to be immeasurable, and the authors also noted little change in influent and effluent nitrogen concentrations, suggesting little if any use for cell synthesis. The protein content of the mixed liquor was only in the range of 20%. A more typical protein content for biomass is 50%, so these results indicate that some inert cell debris was accumulating. The MLVSS concentration varied with time with periods of a steady decline by about 2000 mg/L over 5 to10-day periods (negative yield) followed by a steady increase over a few days. The effluent TSS concentration from the settling chamber was close to 100 mg/L indicating the presence of dispersed growth or accumulation of fine particles.

In laboratory tests with a membrane bioreactor, Choo and Stensel (2000) found low solids yields at an extreme SRT operation. Their reactor was fed a synthetic wastewater consisting of glucose and peptone, and was operated for 350 days at an SRT greater than 50 days. The MLVSS concentration was in the range of 5000-7000 mg/L at 150–250 days of operation. The solids yield during this time was about 0.20 g VSS/g COD_r. From days 250-300, the MLVSS concentration declined steadily from about 6000 to 3000 mg/L, indicating a negative yield condition. The only possible explanation is consumption of biomass by a higher life form, such as protozoan, though no microscopic observations were made. This may also explain the solids reduction changes observed periodically in the extreme SRT reactor operated by Gaudy et al. (1971). Protozoa can consume small particles and perhaps periods of more dispersed solids encouraged their growth and solids consumption. A cyclical pattern is possible in that a high level of dispersed solids encourages protozoan growth and more rapid solids consumption. As the dispersed solids are removed the protozoan activity declines so that the solids consumption rate declines and the solids concentration may increase again.

9.4 Plant Retrofit Considerations

Implementation of an extreme SRT process would require the installation of a membrane separation system to an existing facility. The aeration tank volume may also have to be increased to provide the long SRT required. Figure 9-2 shows the aeration tank HRT as a function of SRT for a system with or without primary treatment using Equation 1.1-1 and MLVSS concentration of 10,000 mg/L. For a 100-day SRT the necessary aeration HRTs are 12 and 24 hours, with and without primary treatment, respectively. Additional aeration equipment also would be necessary to provide the oxygen needed for a greater level of solids reduction. The net solids yields used for this analysis do not account for any significant level of solids consumption by protozoan.

9.5 Assessment of Process

An extreme SRT process would require a membrane bioreactor design, which—coupled with the longer aeration retention time needed—would limit its general use except for relatively small systems.

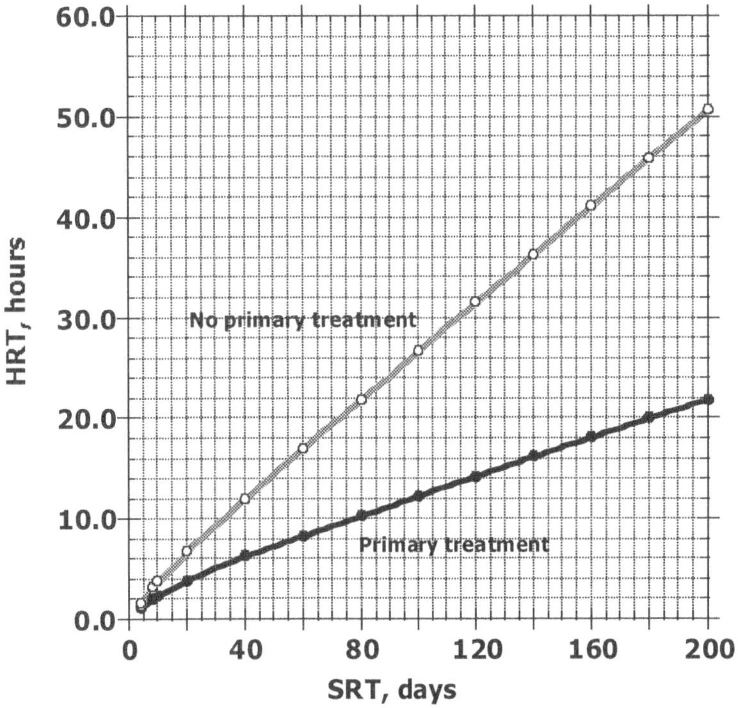

Figure 9-2. Effect of SRT on Required Hydraulic Retention Time (HRT) for Domestic Wastewater Treatment With and Without Primary Treatment and for an MLVSS Concentration of 10,000 mg/L in a Membrane Bioreactor.

This technology might be more cost-effective and attractive for small communities or for on-site treatment for single homes, clusters of homes, and for apartment complexes, if minimal sludge production occurs. For such designs, water reuse also is a consideration in the process selection. A shortcoming of the laboratory work with the extreme SRT operations is that the results were for synthetic wastewater and thus the performance showing minimal sludge production did not account for the effect of nbVSS found in domestic wastewaters. Using Equation 1.2-1 and typical coefficient values for domestic wastewater, the nbVSS in the mixed liquor at a 100-day SRT can account for 50% of the sludge produced.

A key issue regarding a potential advantage for an extreme SRT process is whether conditions are created to encourage proliferation of protozoan to thus result in a biomass reduction due to predation (Chapter 8.0). In that case more can be gained from the extreme SRT operation than just the expected solids loss due to endogenous decay as indicated in Figure 9-1. In effect, the solids yield at a given SRT would also be lower, so that the needed aeration tank volumes indicated by Figure 9-2 may be less.

9.6 Research Needs

Research is needed to determine if an extreme SRT operation can encourage the development of an active protozoan population and thus less biomass production due to predator enhancement. Research on the fate of nbVSS in long SRT systems also would be helpful. The research should address whether these solids can be degraded under the long SRT environment, possibly due to the presence of acclimated bacteria or if they also can be consumed by protozoa under dispersed solids conditions.

CHAPTER 10.0

BENCH-SCALE INVESTIGATION OF COD LOSS IN ANAEROBIC/AEROBIC ACTIVATED SLUDGE TREATMENT

10.1 Experimental Objectives

The literature review summarized in Section 4.2 showed an unexplained COD loss of about 20-30% associated with the A/O (anaerobic/aerobic) and A^2O (anaerobic-anoxic-aerobic) processes. The general consensus was that the incorporation of the anaerobic contact step before aeration is necessary for COD loss and that this type of operation typically will stimulate biological phosphorus removal. The most feasible hypothesis for the COD loss was that fermentation of complex substrates in the anaerobic contact zone results in hydrogen production and evolution in the open tank reactors. For example, in previous studies the addition of acetate to the anaerobic zone did not increase the amount of COD loss. For the previous studies the evidence of COD loss was based on COD balances based on influent and effluent COD and oxygen uptake rate data. There were no cases reported in which volatile solids production was compared between an anaerobic/aerobic and aerobic-only system and shown to be lower.

The initial objective of the bench-scale investigations with A/O reactor systems was to determine if a greater fraction of fermentable substrate added to the anaerobic zone would result in a greater COD loss and thus less VSS production. A second objective that followed results of the first phase of bench scale testing was to determine if VSS production was indeed decreased by the presence of an initial anaerobic contact zone in an activated sludge bench-scale system. In all tests the anaerobic zone was sparged with nitrogen to assure a low oxidation reduction potential (ORP) and no oxidation of COD in the anaerobic zone due to surface transfer of oxygen.

To test the hypothesis that fermentation leads to COD loss in A/O systems, a bench-scale study (West Point Test) was designed to determine if the addition of more fermentable substrate would result in increased COD loss in an A/O process. Two sequencing batch reactors (SBRs) are operated by King County at the Seattle West Point wastewater treatment plant (WWTP) in parallel with identical operating conditions, including solids retention time (SRT), anaerobic and aerobic contact times, total cycle times, number of cycles per day, and influent strength in terms of COD and TSS/VSS concentrations. Both received the West Point WWTP primary effluent and a soluble synthetic organic substrate that almost doubled the influent biodegradable COD concentration. The only difference was that the synthetic substrate addition for Reactor 1 was lactate and for Reactor 2 it was acetate and propionate (2/3 and 1/3, respectively). Lactate was selected as a substrate because it has a known fermentation pathway with hydrogen production. With the A/O operating sequence, both systems were expected to achieve biological phosphorus removal. Propionate was added with the acetate, as some previous work suggests that acetate alone may encourage dominance by a biological population that does not have biological phosphorus removal capability. If fermentation with hydrogen release were occurring with lactate utilization in the anaerobic step, a lower biomass production rate and lower yield in terms of g VSS/g COD removed would be expected for the lactate-fed reactor compared to the acetate-fed reactor.

The results from the West Point testing did not indicate a significant difference in the VSS yield coefficient between the lactate and acetate fed reactors, suggesting that the mechanism of COD loss by evolution from fermentation in the anaerobic zone was not occurring or not at a high enough level to detect a difference in VSS reduction. Based on these findings, it was decided to carry out additional testing at West Point to observe the difference in solids production between the same reactors when one operated with A/O treatment and the other with aerobic treatment only. These tests were done in August 2002 with both reactors fed primary effluent supplemented with lactate. Operating problems occurred during this phase of testing and tests were continued at the University of Washington.

Tests were conducted in the fall of 2002 at the University of Washington environmental engineering laboratory to further determine if a difference in VSS production could be found with A/O treatment versus aerobic treatment only (Test #2). The bench-scale reactor system used was similar to that for West Point tests, including nitrogen stripping during the anaerobic period. Two reactors were operated in parallel, one with the A/O sequence and the other with only aerobic treatment. Both were fed West Point primary effluent wastewater and lactate to assure that sufficient fermentable substrate was available. These tests were done in the lab to provide more controlled conditions for feeding the reactors to assure that both reactors received identical feed. This was done by manually feeding equal volumes of primary effluent supplemented with lactate once per day to sequencing batch reactors (SBRs).

During Test #2, biological phosphorus removal was not maintained in the A/O system. Instead the system was dominated by what has been termed G Bacteria, which are bacteria capable of substrate uptake and storage in the anaerobic zone of an A/O system, but without phosphorus release in the anaerobic zone and excess phosphorus uptake in the aerobic zone, as is characteristic of biological phosphorus removal processes (Cech and Hartman, 1990, 1993;Liu et al., 1996). All of the previous work on COD loss involved A/O processes with biological phosphorus removal. Because no significant difference in VSS production was seen in Test #2, another set of tests were done (Test #3) at the University of Washington laboratory with bench-scale reactors to assure that biological phosphorus removal was occurring before assessing COD loss between an A/O system and aerobic system only. These reactors were operated in parallel using a synthetic wastewater that first contained mainly acetate for the carbon source. Once biological phosphorus removal was established, the feed was adjusted with increasing proportions of lactate in place of acetate.

10.2 Experimental Methods

10.2.1 Reactor Descriptions

West Point Tests

Two bench-scale SBRs were installed at the West Point WWTP and operated by the Seattle King County Department of Natural Resources plant operations staff. As shown in Figure 10-1, each reactor included a magnetic stirrer, air pump and diffuser, feed pumps for primary effluent and synthetic substrate, nitrogen gas supply, and a timer controller. The reactors were constructed of acrylic cylinders and had a nominal operating volume of 8.0 L during aeration. During aeration and the anaerobic contact period, they were mixed with magnetic stirrers and stir bars. Aeration was provided with aquarium air pumps during the aerobic portion of the sequence cycle and controlled with a rotameter and needle valve. Dissolved oxygen (DO) measurements assured that the DO concentration was 2.0 mg/L or greater. During the anaerobic operating phase of the sequencing cycle, the reactors were sparged with nitrogen gas in order to rapidly deplete the mixed liquor of oxygen, which resulted in an immediate drop in redox potential to below –100 mv. The anaerobic sparging also assured that no significant oxygen transfer occurred via the liquid surface so that anaerobic contacting was indeed maintained. A horizontal baffle was also located above the mixed liquor surface to maintain a positive nitrogen blanket above the liquid during the anaerobic contact period. A picture of the West Point reactors arrangement is provided in Figure 10-2.

Test #2: Anaerobic/Aerobic Versus Aerobic Treatment Only With Primary Effluent Feed Supplemented With Lactate

The same reactors used at the West Point site testing were used for Test #2 and both received the same feed, primary effluent plus lactate. The main differences between the West Point and Test #2 experiments are: (1) the feed volume per cycle for Test #2 was 3 L and the full reactor operating volume was 4.0 L, (2) both reactors in Test #2 were operated with only one manual feed cycle per day, (3) one of the reactors (Reactor 2) in Test #2 was operated only under aerobic conditions with no anaerobic step with nitrogen sparging, (4) the reactors were operated with a longer SRT (target SRT of 10 days versus four days), (5) a small flow of carbon dioxide (about 1%) was added with the air sparging flow to help maintain pH below 8.0 in the reactors, and (6) the influent wastewater in Test #2 received a nitrification inhibitor (Allythiourea). Nitrate production was eliminated by using the nitrification inhibitor in Test #2 versus a lower SRT as in the West Point test, so that the presence of nitrate and the effect of nitrate reduction on solids yield would not confound the interpretation of VSS production results.

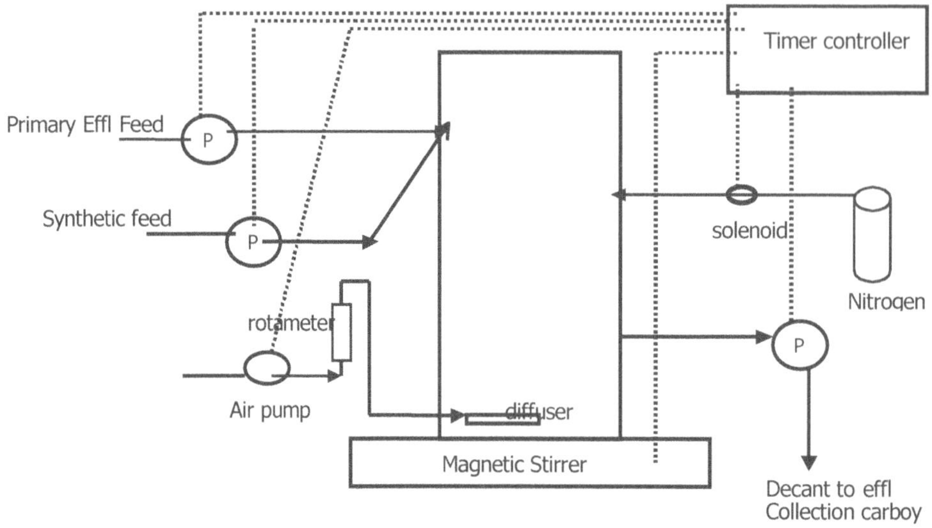

Figure 10-1. SBR Test Reactor Used for Anaerobic/Aerobic Process in COD Loss Experiments at West Point WWTP.

Figure 10-2. Picture of Experimental Set-up at the West Point WWTP.

Test #3: Anaerobic/Aerobic Versus Aerobic-Only Treatment With Synthetic Wastewater

In these tests the system design was similar to that used in Test #2, except a different size and shaped reactor was used. A 2-L Erlenmeyer flask was used for the SBR reactor. The reactors were fed a synthetic wastewater five times per day. A timer controller was used to control the feeding time, nitrogen sparge time, stirrer on/off time, aeration pump on time, and effluent decant pump on time. Nitrogen was sparged through a separate sparge stone in the reactor to evacuate oxygen from the reactor and to develop anaerobic conditions before feeding. A solenoid valve was activated by the timer controller to start and stop nitrogen sparging. A stopper at the top of the reactor had holes in it to provide exit gas flow and a positive pressure in the reactor for the nitrogen headspace during the anaerobic period. Aquarium air pumps were used to provide air through a stone diffuser in the reactor during the aerobic cycles. A small flow of carbon dioxide (about 1%) was also added with the air sparging flow to help maintain pH below 8.0 in the reactors during aeration. Allythiourea was added to inhibit nitrification as in Test #2. Instead of a separate pump for primary effluent and synthetic feed as shown in Figure 10-1, separate pumps were used for inorganic and organic feed solutions of the synthetic feed.

10.2.2 Reactor Operating Conditions

All of the SBR reactors were operated with SRT control with daily manual solids wasting. The amount of daily solids wasting was determined based on daily measurements of the reactor MLSS/MLVSS concentrations at full operating volume, mixed liquor sample volume, effluent volume, and composite effluent TSS/VSS concentrations. The following equations were used to determine the daily manual wasting for the target SRT. Normally the manually wasting rate had to be based on data from the previous day. The actual daily SRT value and ultimately the average SRT for an operating period was calculated from the daily data stored in an Excel spreadsheet.

$$\mathrm{SRT} = \frac{(X)(V)}{Q_W\,(X) = Q_S\,(X) = Q_E\,(X_E)} \qquad 10.2\text{-}1$$

$$Q_W = \frac{V}{SRT} - \frac{Q_E X_E}{X} - Q_S \qquad 10.2\text{-}2$$

Where:

Q_W = Daily manually wasted mixed liquor volume at full SBR depth, L

V = Full SBR volume, L

Q_E = Daily effluent volume, L

Q_S = Daily sample volume, L

X = MLSS concentration at full SBR volume, mg/L

X_E = Effluent composite TSS concentration, mg/L

West Point Tests

From May 2002 and forward, the reactors received primary settled effluent from the West Point WWTP, supplemented with either lactate or a mixture of acetate and propionate. An example of the primary effluent feed characteristics based on data averaged for the month of July is summarized in Table 10-1. During the May to September 2002 steady-state operating periods, the reactor temperature ranged from 15–18 ^{0}C.

Table 10-1. Average Primary Effluent Wastewater Characteristics.

Parameter	Value, mg/L
TCOD	269 ± 27
TSS	59 ± 13
VSS	49 ± 10
NH_4-N	21 ± 9
NO_3-N	0.43 ± 0.31
Sol PO_4-P	3.8 ± 0.7
Total P	4.2 ± 0.8
pH	7.3 ± 0.1 std units

The influent feed to Reactor 1 was supplemented with lactate and the feed to Reactor 2 was supplemented with acetate/propionate by pumping from stock solutions with separate peristaltic pumps controlled by the timer controller during each feed cycle (16 mL per cycle). Supplemental stock solutions with lactate contained from 43,600–48,000 mg/L (with an average of 42,900 mg/L) COD, as determined by standard analysis. Supplemental stock solutions with acetate/propionate contained from 39,000–51,230 mg/L (with an average of 46,200 mg/L) COD as determined by standard analysis. The average contributions of the supplements to the influent COD were 198 mg/L for the lactate reactor and 199 mg/L for the acetate/propionate reactor. Potassium phosphate also was added with the organic carbon supplement at a COD/P ratio of 20:1, so that biological phosphorus removal characteristic of the A/O process was not limited by phosphorus.

The reactors were started in February 2002, but erratic operation occurred in April 2002. A series of operational changes, including sparging with nitrogen during anaerobic contacting, improved operations. By May 2002, a pseudo steady-state operation was achieved with a four-day SRT and supplemental substrate was added at a gradual increasing rate. By June EBPR was achieved and a stable operation was maintained. In July the same SRT was maintained and the supplemental organic substrate concentrations were increased. In August 2002, the operation was modified to study the solids production for an A/O system versus an aerobic-only system due to the lack of apparent COD loss when comparing the effect of adding a fermentable and non-fermentable supplemental COD to the two A/O systems operated in parallel.

The operating sequence for the SBRs is shown in Table 10-2 for the operating conditions with an A/O process and with an aerobic process only. For the parallel reactor operations, the aerobic process was held in an idle mode after settling and decanting while the A/O process went through its feeding and anaerobic contact period. The aerobic only system was fed just near the end of the anaerobic contact period.

Table 10-2. SBR Cycle Schedule (5 cycles per day) for Anaerobic/Aerobic and Aerobic Only Process Operating Conditions.

Sequence Step	Time, h	
	Anaerobic/aerobic	Aerobic only
Fill	0.05	0.05
Anaerobic contact	0.75	—
Aeration	3.0	3.0
Settle	0.50	0.50
Decant	0.50	0.50
Idle	—	0.75
Total cycle time	4.80	4.80

A Chrontrol® timer automatically controlled the SBR operation. During the fill period, the timer turned on the primary effluent and synthetic organic feed peristaltic feed pumps for five minutes. Just before the fill, the timer turned on the magnetic stirrer and opened the solenoid on the nitrogen gas feed line so that the reactor was flooded with nitrogen. Nitrogen flooding of the reactor continued during the anaerobic period. After the anaerobic contact period, the magnetic stirrer was kept on, the nitrogen gas solenoid valve was closed, and the air pump was turned on. For the decant and settle periods, the air pump and magnetic stirrer were turned off. During the decant period the controller turned on an effluent withdrawal peristaltic pump. The reactors were operated with SRT control and the daily target SRT was set at four days. Manual wasting of reactor mixed liquor was done Monday through Friday, based on daily measurements of mixed liquor and effluent TSS concentrations, and effluent and sampling volumes. For weekend days (Saturday and Sunday), a preset volume was wasted from the reactors daily using a timer and peristaltic pump, with the waste volume based on the weekday wasting rates.

The SBR cultures were seeded with activated sludge from the West Point WWTP pure oxygen activated sludge basins. Once anaerobic conditions were firmly established for the anaerobic contact period, biological phosphorus removal peaked out during a two-week period.

Test #2: Anaerobic/Aerobic Versus Aerobic Treatment Only With Primary Effluent Feed Supplemented With Lactate

In view of the findings from the West Point tests, Test #2 was done to evaluate COD loss in the A/O process by comparing its VSS production to that for an aerobic treatment process only. The reactors were seeded with mixed liquor from the East Bay Municipal Utility District, Oakland, CA, activated sludge facility that was operated with an A/O process and was accomplishing biological phosphorus removal. Both received West Point primary effluent wastewater plus additional soluble fermentable COD in the form of lactate. Thus, Reactor 1 continued to operate with the A/O process and Reactor 2 was operated without the anaerobic contact period. The cycle times for the two reactors are given in Table 10-3.

When Reactor 1 was in the anaerobic contact period, Reactor 2 remained in an idle period after settling and decanting. Reactor 2 was fed just before Reactor 1 started the aerobic period and both reactors were operated with the same aeration time and 10-day SRT. ORP measurements of less than –100 mV during the anaerobic contact phase assured anaerobic conditions.

Table 10-3. SBR Cycle Schedule (5 cycles per day) for Test #2.

Sequence Step	Time, h	
	Reactor 1 (Anaerobic/Aerobic)	Reactor 2 (Aerobic)
Fill	0.10	0.1
Anaerobic contact	0.90	-
Aeration	22.0	22.0
Settle	0.75	0.75
Decant	0.25	0.25
Idle		0.9
Total cycle time	24.0	24.0

The feed was prepared manually each day. A lactate stock solution containing bicarbonate buffer and potassium phosphate was added at 45 mL to a feed container filled to a total of 6.0 L with West Point primary effluent wastewater. The feed was well mixed and equal portions of 3.0 L each were added to Reactors 1 and 2. The lactate stock solution consisted of 40 mL/L 85% lactic acid, 20 g/L $NaHCO_3$, and 10 g/L KH_2PO_4. The lactate solution provided 272 mg/L of additional COD to the feed and represented about one-half of the total feed COD. The additional phosphorus provided to the feed by the potassium phosphate in the stock lactate solution was 17 mg/L based on the total feed volume.

Test #3: Anaerobic/Aerobic Versus Aerobic Only Treatment with Synthetic Wastewater

For Test #3, two SBRs were operated in parallel (A/O versus aerobic) with synthetic feed. The synthetic feed composition based on the feed flow is summarized in Table 10-4. The organic and inorganic feed solutions were prepared in separate containers and fed separately by peristaltic pumps to the reactor during each feed cycle. The total feed volume per cycle was 1.0 L, with 250 mL and 750 mL from the organic and inorganic feed reservoirs, respectively.

Both reactors were operated in an environmental chamber with temperature control at 15 °C. The target SRT was eight days. The sequence times for each cycle are summarized in Table 10-5. During the time that the anaerobic reactor was sparged with nitrogen, feed, and mixed for the anaerobic contact time, the aerobic reactor remained in an idle mode. The reactors were operated with equal aeration times and SRTs. The aerobic-only SBR was filled at the end of the anaerobic contact period for Reactor 1.

Table 10-4. Synthetic Feed Composition for Test #3 (Prepared with De-Ionized Water).

Constituent	Concentration, mg/L	Constituent	Concentration, mg/L
Sodium Acetate[a]	400	$FeCl_3$-$6H_2O$	0.5
Yeast Extract	5	H_3BO_3	0.04
Peptone	5	$CuSO_4$-5 H_2O	0.009
Est. Total COD	200	KI	0.005
$NaHCO_3$	250	Na_2MoO_4-$2H_2O$	0.018
NH_4Cl[b]	45.9	$ZnSO_4$- $7H_2O$	0.036
KH_2PO_4-$2H_2O$[c]	50.0	$CoCl_2$- $6H_2O$	0.045
KCl	45	$MnCl_2$- $4H_2O$	0.036
$MgSO_4$-$7H_2O$	180	$Al_2(SO_4)_3$- $18H_2O$	0.162
$CaCl_2$-$2H_2O$	140	$NiCl_2$- $6H_2O$	0.012
		EDTA	6.0

[a]188 mg COD/L , [b]12 mg/L as N, [c]11.7 mg/L as P

Table 10-5. SBR Cycle Schedule (5 cycles per day) for Test #3.

Sequence Step	Time, h	
	Reactor 1 (Anaerobic/Aerobic)	Reactor 2 (Aerobic)
Nitrogen Sparge	0.20	-
Fill	0.10	0.10
Anaerobic contact	0.80	-
Aeration	2.7	2.7
Settle	0.75	0.75
Decant	0.25	0.25
Idle	-	1.0
Total cycle time	4.8	4.8

10.2.3 Sampling and Analyses

West Point Tests

Daily composite influent and effluent samples were analyzed for total COD, total and volatile suspended solids (TSS/VSS), ammonium nitrogen, nitrate nitrogen, soluble orthophosphorus, and total phosphorus. Orthophosphate also was determined on filtered grab samples taken toward the end of the anaerobic period to confirm the presence of phosphorus release during the anaerobic contacting period. Mixed liquor grab samples were analyzed for MLSS and MLVSS, total COD. During steady-state testing, influent and effluent SO_4 analyses were done.

In order to demonstrate the effectiveness of the anaerobic period, redox levels were determined in the anaerobic period, usually toward the end of the period, using a Broadley James oxidation-reduction potential (ORP) electrode. Typical results are shown in Table 10-6. The low redox values clearly indicate the anaerobic nature of this portion of the SBR cycle and the high levels of phosphate in the mixed liquor compared to the influent phosphate levels showed that phosphorus release was occurring in the anaerobic step as expected for the excess biological phosphorus uptake process. This conclusion was further supported by the low effluent orthophosphorus concentrations (0.2 ± 0.1 as P) for both SBRs.

Table 10-6. Redox Levels and Phosphorus Release in the Anaerobic Period of the SBR Operation.

Reactor	ORP mV	Ortho Phosphorus mg P/L
Lactate	-97.6 ± 44.8	45.7 ± 9.6
Acetate/Propionate	-106.1 ± 47.4	43.7 ± 9.2

DO concentrations and pH were monitored daily in the reactors. DO concentrations were typically in the 1.0–4.0 mg/L range, with the DO concentrations lowest in the first 30 min after feeding and increasing with time of aeration to about 4.0 mg/L at the end of the aeration period.

At steady-state conditions and towards the end of the reactor operation, oxygen uptakes rates (OUR) were determined during the aeration portion of an SBR cycle for three different days. On days when OUR tests were done, a primary effluent grab sample was obtained and used for manual feeding of each reactor for the feed cycle in which the OUR test was done. OURs were done for one reactor first and then the other in the subsequent cycle. The portion of feed for the second reactor was stored until the feed cycle for its OUR test. The organic supplement of lactate or acetate/propionate was added with the primary effluent for the manually fed cycle. In this way each reactor received the same feed COD for the cycle used for the OUR test.

The OUR test started at the beginning of the aeration period and continued over the entire aeration period, so that the amount of oxygen consumed in a feed cycle could be determined. A Yellows Springs Instrument (YSI) DO probe was placed in the reactor before the test. The DO concentration was recorded with time using digital readouts from the YSI meter. During the OUR testing, the mixing was maintained and aeration was applied intermittently to rapidly raise the DO concentration to 5.0–6.0 mg/L, within one to two minutes, after it was depleted by the biological respiration in a linear fashion to 1.0–2.0 mg/L. During the time without aeration, the decline in the DO concentration with time was monitored to determine the oxygen uptake rate. The slope of the DO concentration change with time was the OUR at that time in the aerobic period. In this way a graph of the OUR versus time (mg/L –hr) could be obtained for the entire aeration period. The area under the curve was calculated to yield the total oxygen consumption for the aeration cycle. The carbonaceous oxygen consumption was then calculated by subtracting the oxygen used for nitrification based on nitrate and nitrate analyses of the reactor contents before and after the aeration period.

Test #2: Anaerobic/Aerobic Versus Aerobic Treatment Only With Primary Effluent Feed Supplemented With Lactate.

As in the West Point tests, daily composite influent and effluent samples were analyzed for total COD, TSS/VSS, ammonium-nitrogen, nitrate-nitrogen, and soluble orthophosphorus. Orthophosphate was also determined periodically on filtered grab samples taken toward the end of the anaerobic period to confirm the presence of phosphorus release during the anaerobic contacting period. The ORP was checked periodically during the anaerobic period to assure anaerobic conditions. Mixed liquor grab samples were analyzed daily for MLSS and MLVSS.

During Test #2, it was not possible to sustain biological phosphorus removal in the A/O process, in spite of a low ORP indicating anaerobic conditions during the anaerobic contact period. The soluble COD concentration was monitored following fill through the anaerobic and aerobic periods and indicated that most of the soluble COD was consumed during the anaerobic contact period in spite of no or little phosphorus release. This behavior is indicative of a system dominated by G bacteria. Microscopic analysis at 1000 magnification with a phase contrast microscope confirmed that a high level of G bacteria were present, as indicated by the characteristic tetrad cocci bacteria which were easily observed throughout the mixed liquor. Thus, although COD was removed in the anaerobic period, this removal apparently was accomplished by G bacteria rather than by phosphorus-removing bacteria.

Test #3: Anaerobic/Aerobic Versus Aerobic-Only Treatment With Synthetic Wastewater.

Test #3 was made possible by the availability of two exchange students from Sichuan University in China. A number of months were required to train the students on the operation of an SBR system for biological phosphorus removal. By April 2003, they had developed a consistent biological phosphorus removal operation and properly maintained the parallel A/O and aerobic-only SBR systems treating synthetic wastewater. For the first phase of this effort, only the A/O system was operated until efficient biological phosphorus removal was demonstrated. The reactor was seeded with mixed liquor from the Durham wastewater treatment facility in Portland, OR, which was achieving a high level of biological phosphorus removal. During the anaerobic period, the ORP, phosphorus release (increase in orthophosphorus concentration), and soluble COD uptake were monitored with time. A successful operation was indicated by a phosphorus release to COD uptake ratio close to theoretical values (0.4-0.5 g/g), low ORP in the anaerobic period, and a low effluent soluble phosphorus concentration (<0.5 mg/L). Once good biological phosphorus removal was established, the reactor contents were split between two SBRs; one with the A/O and the other with the aerobic-only operation. After one month of operating with acetate as the primary COD substrate, the feed was modified to 50% lactate and 50% acetate and then finally to 100% lactate on a COD basis to promote fermentation in the anaerobic period.

The SBR system was monitored in the same way as that for Test #3, except that the influent composition was known based on the feed preparation. Daily measurements were made on composite effluent TSS/VSS, soluble COD, and soluble phosphorus concentrations, and reactor MLSS/MLVSS, DO concentrations, and pH.

10.2.4 Analytical Procedures

10.2.4.1 Ammonia Nitrogen

An adaptation of the Salicylate Method (APHA, 1998) was followed for NH_4-N measurements, using a reagent prepared by the Hach Chemical Co. (Loveland, CO). Samples were filtered through a 0.45 μm filter and diluted to give a concentration in the range of 0–50 mg N/L. A 0.1 mL filtered sample was pipetted to a high range (0–50 mg/L NH_4-N) 10-mL reagent vial and then 0.1 mL deionized water was prepared for another reagent vial, which served as a blank. The vials received the contents of one ammonia salicylate and one ammonia cyanurate reagent powder pillow then left for 20 minutes for the reaction. Color development was measured by absorbance at 655 nm using a Hach DR/4000 Spectrophotometer, and the blank vial was used to zero the instrument. The NH_4-N concentration of the unknown sample was obtained from a standard curve of absorbance versus NH_4-N concentration. The standard curve had 5, 10, 20, 30, and 40 mg/L concentration values and was done in duplicate. The linear fit of absorbance versus concentration had a correlation (R^2) greater than 0.99. Samples with NH_4-N concentrations greater than 50 mg/L were diluted in a volumetric flask with deionized water.

10.2.4.2 Nitrate Nitrogen

Nitrate nitrogen was determined using Hach Nitrate high range (0–30 mg/L NO_3-N) Reagent Vials (Hach Chemical Co.) and the chromotropic acid method, with 1.0 mL filtered samples pipetted into 10-mL reagent vials. Vials received the contents of one NitraVer X reagent powder pillow and were inverted 10 times for mixing. After a five-minute unstirred reaction period, the absorbance was determined at 410 nm using a HACH DR/4000 Spectrophotometer. The NO_3-N concentration of the unknown sample was obtained from a standard curve of absorbance versus NO_3-N concentration. The standard curve was prepared with 2, 5, 10, 20, and 30 mg/L concentrations done in duplicate. The linear fit of the absorbance versus concentration curve had a correlation (R^2) greater than 0.99. Nitrite is a positive interference in this test at concentrations above 12 mg/L. Samples were diluted with deionized water in a volumetric flask whenever NO_3-N or NO_2-N concentration were greater than 30 mg/L and 12 mg/L, respectively.

10.2.4.3 Nitrite Nitrogen

For samples with NO_2-N concentration less than 5 mg/L, the Hach Nitrite Low Range (0–0.5 mg/L NO_2-N) diazotization method was used. Samples were filtered through a 0.45 µm filter and diluted to give a concentration in the range of 0–0.5 mg N/L. A 5.0 mL filtered sample was pipetted to a 25-mL sample cell. A Hach NitriVer 3 Nitrite LR reagent powder pillow was added to the sample cell, which was shaken to dissolve the powder, and then left 20 minutes for reaction. The absorbance of the sample was then read on a Hach DR/4000 Spectrophotometer at 507 nm. For samples with NO_2-N concentrations greater than 5 mg/L, an adaptation of the Hach Nitrite High Range (0–250 mg/L NO_2-N) ferrous sulfate method was followed. In this method, 10 mL of filtered sample was added to a 25-mL sample cell, which received the content of one NitriVer 2 Nitrite HR Reagent powder pillow. After shaking to dissolve the powder, the sample cell was left for 10 minutes for reaction. The absorbance of the sample was then read on a Hach DR/4000 Spectrophotometer at 585 nm. A NO_2-N standard curve was developed using concentrations of 10, 20, 50, 100, and 200 mg/L done in duplicate. A linear fit of the absorbance versus concentration curve had a correlation (R^2) greater than 0.99.

10.2.4.4 Chemical Oxygen Demand

Chemical oxygen demand (COD) was measured by digesting samples in Hach Low Range (0–150 mg/L) COD Digestion Reagent Vials (Hach Chemical Co.) and then reading the absorbance of the samples to indicate reaction with potassium dichromate, and thus the level of COD. A 2.0 mL sample was pipetted to a Hach COD reagent vial, digested at 150 °C for two hours in a Hach COD reactor Model 16500-10 heater block, and allowed to cool to the touch. The absorbance of the sample was then read on a Hach DR/4000 Spectrophotometer at 620 nm. A COD calibration curve was developed using concentrations of 25, 50, 75, 100, and 125 mg/L of a COD standard prepared by adding 425 mg of potassium biphthalate to 1.0 L deionized water and diluting to the standard solution. Samples with a COD concentration higher than 150 mg/L had to be diluted with deionized water. For samples requiring dilution to one-half the concentration, the vial received 1 mL of sample and 1 mL of deionized water. Higher concentrations required dilution in a volumetric flask. For soluble COD, the sample was filtered with a syringe filter containing a 0.45 μm membrane filter.

10.2.4.5 Total Suspended Solid and Volatile Suspended Solids

Total and volatile suspended solids (TSS/VSS) were measured according to Standard Methods 2540 D and C (APHA, 1998). A known volume of the sample was filtered through a glass fiber filter. The filter was dried for at least one hour in a 105 °C oven, cooled in a desiccator and weighed. The weight increase on the filter was due to the TSS. The filter was then placed in a 550 °C oven for 15 minutes, cooled and weighed. The weight loss on the filter was due to the VSS. The filters were prepared by vacuum washing them three times with deionized water and placed in the 550 °C oven to remove any volatile solids present on the filter paper. The accuracy of the method is about 10% for concentrations greater than 200 mg/L (APHA, 1998).

10.2.4.6 Soluble Orthophosphate

Soluble orthophosphate concentrations were determined by using the Vanadomolybdo phosphonic acid colorimetric method (4500-P.C. in APHA, 1998). A wavelength of 430 nm was used to measure absorbance on the Hach spectrophotometer after color development. The sensitivity of the test method is at an orthophosphate concentration as low as 0.05 mg/L. Standard curves for absorbance versus concentration were developed with potassium phosphate at concentrations of 0.05, 0.20, 0.40, 0.60, 0.80, and 1.00 mg/L as P.

10.2.4.7 Dissolved Oxygen, Temperature and pH

DO concentrations were measured using a YSI meter and YSI model 5750 DO probe calibrated at zero and saturated concentrations. For the zero DO calibration, the DO probe was placed in a sodium sulphite solution, held until observing a DO reading stable and adjusting the meter DO reading to zero. The DO probe was then placed in a beaker of oxygen saturated water. Using data from a table of oxygen saturation concentration versus temperature (Tchobanoglous et al., 2003), the DO concentration for the meter calibration was set. Temperature was recorded from a thermometer with 0.2 °C temperature divisions. A Beckman model #10 pH meter equipped with a Corning General Purpose Combination pH probe (Probe#476530) was used to measure pH. The pH meter was calibrated at pH 4, 7, 10 before each pH measurement.

10.2.4.8 Total P and Sulfate

The analyses for Total P and sulfate concentrations were performed by the West Point WWTP lab staff using procedure described by APHA (1998).

10.3 Results

10.3.1 West Point Tests

Experimental results for the West Point tests to determine the effect of adding a fermentable substrate to the A/O process are presented first. The reactors were started in February 2002, but stable operating conditions were not developed at the test site until May. A series of process operating modifications were made to ensure that anaerobic conditions were maintained for the appropriate time in each feed cycle of the SBR operation. An important operating change was the addition of nitrogen flushing during the anaerobic phase to ensure that anaerobic conditions occurred, and no aerobic degradation was occurring due to surface transfer of oxygen. By June a steady state operation was achieved for the reactors based on having a time period equal to at least three SRTs with stable uniform operation (12 days). Stable operation continued through June, July, and August.

A comparison of the influent characteristics and measured effluent values are presented first for Reactors 1 and 2 before summarizing solids production results. These results are averaged with standard deviations for June and July separately. The primary effluent feed was sampled separately by a composite sampler before adding the supplemental COD, so its characteristics are provided in Tables 10-7 and 10-8 for June and July, respectively, for total COD (TCOD), TSS/VSS, NH_4-N, total phosphorus, and soluble phosphorus concentrations. The actual flow fed daily to each reactor was determined by measuring the volume in the effluent collection container each day plus the volume removed from the reactor for sampling and solids wasting. As shown, the volumes fed to each reactor were not exact but close. This close monitoring of the reactor feed volume was important for the COD mass balance and solids yield calculations.

The amount of supplemental COD fed each day to the reactors was determined by daily changes in the supplemental feed reservoir volume and knowing its COD concentration. For June the additional COD provided to the influent by lactate for Reactor 1 (R1) and acetate/propionate for Reactor 2 (R2) was 173 and 144 mg/L, respectively. This was a 40–50% increase in the influent rbCOD available for biological phosphorus accumulating organisms (PAOs). For July the additional rbCOD provided by lactate to R1 and acetate/propionate to R2 averaged 197 and 172 mg/L, respectively.

Table 10-7. Average SBR Influent and Effluent Characteristics for June, 2002 During West Point Tests. Reactor 1 Feed Supplemented with Lactate and Reactor 2 Feed Supplemented with Acetate/Propionate.

Parameter	Influent Reactor 1	Influent Reactor 2	Effluent Reactor 1	Effluent Reactor 2
Flow, L/d	19.6 ± 1.3	20.8 ± 1.4		
PE TCOD, mg/L	342 ± 83	342 ± 83		
TCOD with supplement, mg/L	515 ± 46	486 ± 48	51± 17	52 ± 22
TSS, mg/L	83 ± 29	83 ± 29	19± 7	18 ± 7
VSS, mg/L	70 ± 25	70 ± 25	14 ± 5	14 ± 6
NH_4-N, mg/L	23 ± 4	23 ± 4	3.4 ± 3	4.2 ± 5
NO_3-N, mg/L			5.4 ± 2	5.7 ± 2
Sol PO_4-P, mg/L	12.7 ± 0.8	12.7 ± 0.8	0.3 ± 0.3	0.3 ± 0.3
Total P, mg/L	13.1 ± 0.7	13.1 ± 0.7	0.6± 0.4	0.5± 0.3
pH			7.2±0.1	7.2±0.1
Average Temperature, ^{0}C			16.0	16.0

For both June and July the reactors were maintained at the 4-day SRT operation, and their effluent characteristics are strikingly similar. The average effluent TCOD and TSS/VSS concentrations are within 1–2 mg/L for R1 and R2 for June and July. Both reactors showed similar levels of nitrification. A similar and higher average effluent NH_4-N concentration in the range of 3.4–4.2 mg/L NH_4-N is shown for June and an average value in the range of 1.2–1.4 mg/L is shown for July. The higher average reactor temperature of 17.5 °C in July versus 16 °C in June can account for the higher nitrification efficiency. The average effluent NO_3-N concentrations were also similar for the two reactors and agree with the expected amount of nitrate reduction during the anaerobic contact period.

Table 10-8. Average SBR Influent and Effluent Characteristics for July 2002 During West Point Tests. Reactor 1 Feed Supplemented with Lactate and Reactor 2 Feed Supplemented with Acetate/Propionate.

Parameter	Influent Reactor 1	Influent Reactor 2	Effluent Reactor 1	Effluent Reactor 2
Flow, L/d	19.1 ± 1.6	19.4 ± 5		
PE TCOD, mg/L	269 ± 26	269 ± 26		
TCOD with supplement, mg/L	466 ± 31	441± 25	38± 4	40 ± 8
TSS, mg/L	60 ± 12	60 ± 12	14± 6	16 ± 10
VSS, mg/L	50 ± 10	50 ± 10	10 ± 5	12 ± 7
NH_4-N, mg/L	23 ± 7	23 ± 7	1.4 ± 3	1.2 ± 2
NO_3-N, mg/L			5.8 ± 1	6.0 ± 1
Sol PO_4-P, mg/L	13.9 ± 0.8	13.9 ± 0.8	0.2 ± 0.1	0.2 ± 0.1
Total P, mg/L	14.4 ± 0.7	14.4 ± 0.7	1.1 ± 0.3	1.0 ± 0.5
pH			7.3 ±0.1	7.2 ±0.1
Average Temperature, ^{0}C			17.5	17.5

Both reactors achieved similar and excellent biological phosphorus removal with effluent PO_4-P concentrations averaging 0.3 and 0.2 mg/L for June and July, respectively. Daily ORP and phosphorus concentration measurements during the anaerobic contact period confirmed good anaerobic conditions (ORP from -100 to -150 mV) and biological phosphorus removal activity with phosphorus release causing orthophosphorus concentrations to typically range from 30–60 mg/L as P.

The average solids production and solids yield results (g VSS/g COD removal) are summarized in Tables 10-9 and 10-10 for June and July, respectively. The average daily values for COD removal (influent COD minus effluent soluble COD) and VSS production for each month are summarized. The VSS production values equal the total amount of solids wasted manually each day for SRT control, plus the effluent solids and the solids wasted via sampling. The waste solids and sample solids concentrations were the daily MLVSS concentrations. The average amount of COD removed daily for R1 and R2 was similar. R2 had slightly lower average influent COD concentrations, but a slightly higher average flow resulting in a similar COD loading as R1 for June and July. For both months the average MLVSS concentrations were similar and almost identical at 1709 and 1667 mg/L MLVSS for R1 and R2 in June and 1768 and 1744 for R1 and R2 in July, respectively. The yield was calculated as the ratio of the mg VSS produced daily divided by the mg COD removal daily. The average observed yields were almost identical for both reactors also at 0.37 g/g for R1 and R2 in June and 0.44 and 0.43 for R1 and R2, respectively, in July. The observed yields are within the range of yields reported in the literature for activated sludge treatment of primary effluent. The average higher observed yield value for July versus June is likely a result of wastewater changes, with possibly a higher nonbiodegradable solids concentration in the primary effluent during July.

Fundamental process design considerations for the activated sludge process hold that the solids production and MLVSS concentrations should be equal at equal SRT, aeration volume or aeration time (for SBRs), and applied COD and suspended solids loading, unless there is some unique difference in the process design or operation of one of the systems. A comparison of the MLVSS concentrations for systems R1 and R2 show similarities in their MLVSS concentrations during June and July (Figure 10-3), but with obvious scatter in their comparisons with either R1 or R2 values higher on different days. The comparison is more readily observed by a plot of the R1:R2 MLVSS ratio in Figure 10-4. If significant COD loss was occurring in system R1 due to lactate fermentation, the number of date points with R1/R2 less than 1.0 should be more than half the data points. In this case there are actually more data points with R1/R2 greater than 1.0. The variability, typically ± 10%, is due to sampling and analysis within the capability of TSS/VSS measurements.

Table 10-9. Average Solids Yield Values for SBRs for June 2002 During West Point Tests. Reactor 1 Feed Supplemented with Lactate and Reactor 2 Feed Supplemented with Acetate/Propionate.

Parameter	Reactor 1	Reactor 2
COD removed*, mg/d	9243 ± 1140	9198 ± 1243
MLVSS, mg/L	1709 ± 200	1667 ± 183
MLSS, mg/L	2250 ± 306	2162 ± 313
MLVSS/MLSS	0.76 ± 0.03	0.76 ± 0.03
SRT, days	4.0	4.0
VSS production, mg/d	3454 ± 415	3318 ± 319
Yield, g VSS/g CODr	0.37 ± 0.05	0.37± 0.05

*Influent TCOD minus effluent soluble COD

Table 10-10. Average Solids Yield Values for SBRs for July 2002 During West Point Tests. Reactor 1 Feed Supplemented with Lactate and Reactor 2 Feed Supplemented with Acetate/Propionate.

Parameter	Reactor 1	Reactor 2
COD removed*, mg/d	8184 ± 707	8152 ± 626
MLVSS, mg/L	1768 ± 173	1744 ± 179
MLSS, mg/L	2394 ± 223	2374 ± 243
MLVSS/MLSS	0.74 ± 0.02	0.74 ± 0.02
SRT, days	4.0	4.0
VSS production, mg/d	3573 ± 322	3452 ± 360
Yield, g VSS/g CODr	0.44 ± 0.05	0.43 ± 0.05

*Influent TCOD minus effluent soluble COD

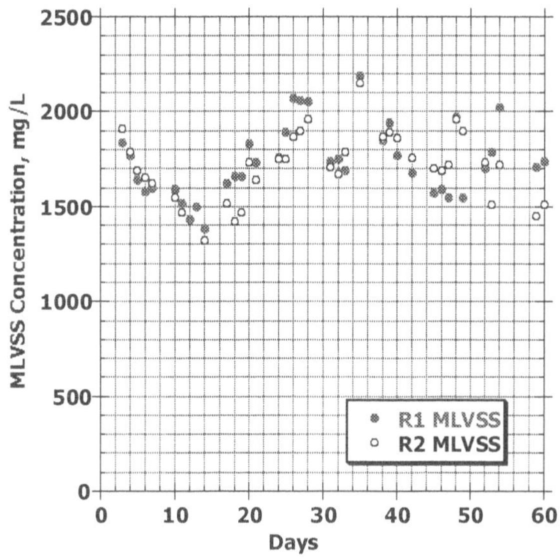

Figure 10-3. MLVSS Concentrations Versus Time for Reactors 1 and 2 During June to July, 2002 West Point Tests. Reactor 1 Supplemented With Lactate COD and Reactor 2 Supplemented With Acetate COD.

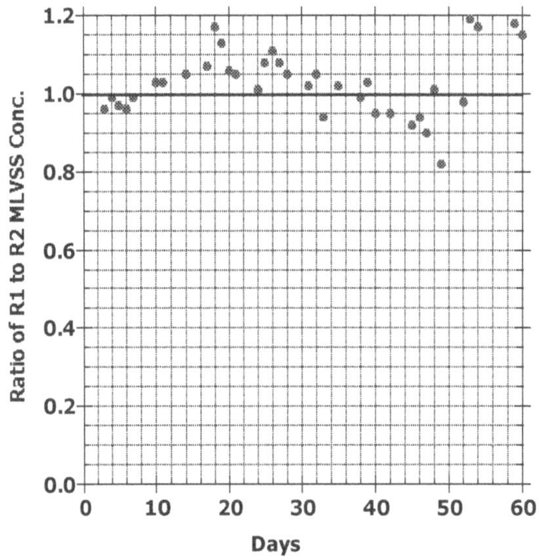

Figure 10-4. Ratio of MLVSS Concentrations Versus Time for Reactors 1 and 2 During June to July, 2002 West Point Tests. Reactor 1 Supplemented With Lactate COD and Reactor 2 Supplemented With Acetate COD.

Another parameter that theoretically should reflect differences in process performance, if COD loss is occurring, is the concentration of orthophosphate in the anaerobic contact period. Phosphorus is released by PAOs in proportion to the uptake of volatile fatty acids. As the PAOs consume COD in the anaerobic zone there is a stoichiometric release of orthophosphorus related to energy utilization for substrate uptake and the formation of carbon storage products. The measured orthophosphorus concentrations for R1 and R2 during June and July are summarized in Figure 10-5, and their ratios are given with time in Figure 10-6. Again, no clear trend is evident to suggest a COD loss in the anaerobic zone as there are almost an equal number of points with R1/R2 greater than 1.0 and less than 1.0.

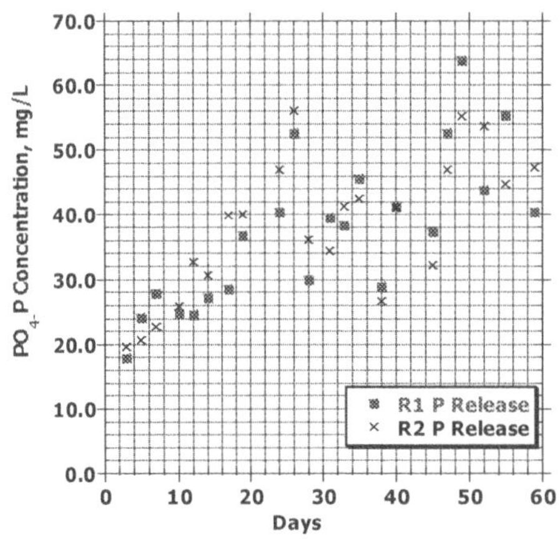

Figure 10-5. Anaerobic Period Final Orthophosphate Concentrations Versus Time for Reactors 1 and 2 for Period of June 1 to July 31, 2002 (Day 0 is June 1). Reactor 1 has Lactate Supplemental COD and Reactor 2 has Acetate/Propionate Supplemental COD.

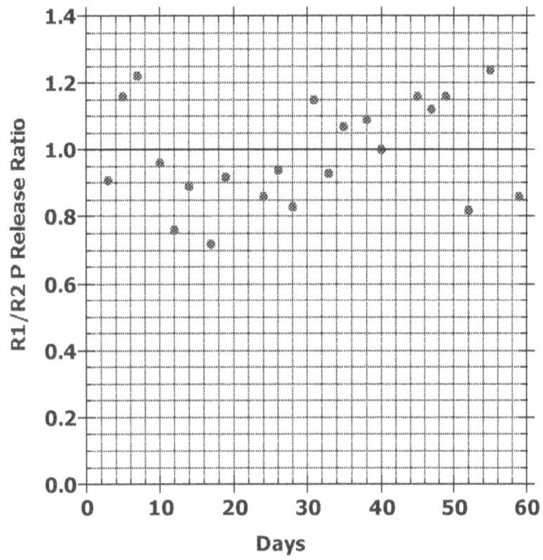

Figure 10-6. Ratio of Phosphorus Release in Anaerobic Period for Reactor 1 Versus Reactor 2 During June to July, 2002 West Point Tests. Reactor 1 Supplemented with Lactate COD and Reactor 2 Supplemented with Acetate COD.

In addition to monitoring phosphorus concentrations across the A/O process periods, profiles done in late July also measured sulfates and oxidized nitrogen. As expected the oxidized nitrogen was quickly removed in the anaerobic period. The changes in sulfate concentration from the start and end of the anaerobic period, shown for profiles on seven different days in July (Table 10-11), are negligible and indicate that no sulfate reduction is occurring in the anaerobic step. Thus, sulfate is not serving as an electron acceptor to account for COD consumption in the anaerobic period. The exposure to oxygen in the aerobic period likely inhibits the development of a sulfate-reducing bacteria population.

Table 10-11. Sulfate Concentrations (mg/L) During SBR Cycle.

DATE	REACTOR	Sulfate Concentrations, mg/L		
		Anaerobic Start	Anaerobic End	Aerobic End
07/17/02	R1	53	50	53
	R2	53	58	58
07/19/02	R1	41	43	41
	R2	39	41	41
07/22/02	R1	48	44	41
	R2	44	44	45
07/25/02	R1	52	55	50
	R2	52	51	56
07/29/02	R1	46	49	48
	R2	45	45	45
07/30/02	R1	49	48	48
	R2	48	48	48
07/31/02	R1	41	44	45
	R2	48	44	45

Oxygen Uptake Rate Tests

Following batch feeding and the anaerobic contact period, the OUR of the reactor mixed liquor was measured for the entire aeration period (180 min). An example of an OUR test result is shown in Figure 10-7. Data points are shown at incremental times and represent the average OUR over a time period in which the reactor DO concentrations were raised by aeration and then declined during the OUR test during mixing without aeration. The data time point represents the SBR aeration time that elapsed after the time used to measure the OUR.

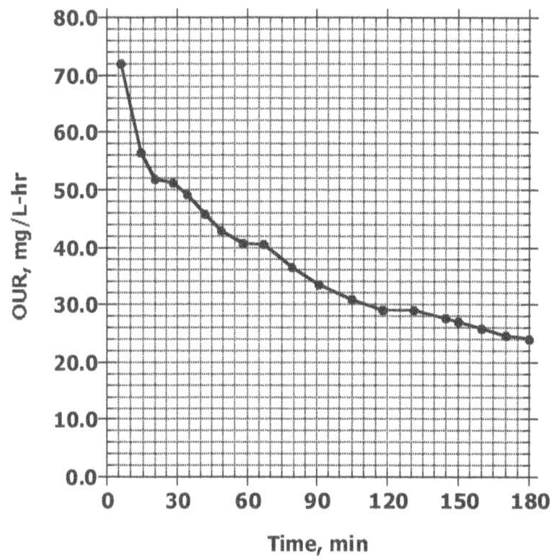

Figure 10-7. Example Oxygen Uptake Rate Test Results Showing OUR Versus Time for Aeration Period. Reactor 1 with Lactate COD Supplement.

The area under the OUR versus time curve represents the amount of oxygen consumed in mg during the aeration period (OC_{our}). The OC_{our} was calculated for each test by multiplying the incremental time between data points by the average OUR value measured for the two incremental data points. For the example plot, the OC_{our} for the aeration period is 908 mg.

A comparison of the oxygen consumption for COD removal between R1 and R2 may be used to determine if there was a COD loss due to supplemental feeding of a fermentable substrate (lactate) into R1 versus acetate/propionate for R2. Assuming equal biomass yields for the equal SRT operation, a lower ratio of carbonaceous oxygen consumed per g COD applied (oxygen consumption ration-OCR) would be expected for an operation with COD loss. The OCR was determined by adding to the OUR data the equivalent amount of oxygen provided by nitrate and nitrite reduction during the anaerobic period and subtracting the oxygen used for nitrification during the aerobic period. For each OUR test, the total COD of the batch feed was determined, and the nitrate and nitrite concentrations were determined in the mixed liquor at the start and end of the anaerobic period and at the end of the aerobic period. The carbonaceous oxygen consumption calculation is described by the following equations and is identical to the method used in the COD balance equations by Randall et al. (1992) to determine COD loss based on OURs, influent and effluent CODs, and changes in oxidized nitrogen.

$$OC_{CARB} = OC_{OUR} + OC_{DN} - OC_{NIT} \qquad 10.3\text{-}1$$

Where:

OC_{CARB} = Oxygen consumption in aerobic period for COD removal, mg

OC_{OUR} = Measured oxygen consumption during aeration period from OUR data, mg

OC_{DN} = Equivalent amount of oxygen used during anaerobic period for nitrate and nitrite reduction, mg

OC_{NIT} = Oxygen consumed during aeration period for nitrification, m.

$$OC_{DN} = [2.86\,(\Delta NO_3)_{AN} + 1.71\,(\Delta NO_2)_{AN}]\,V \qquad 10.3\text{-}2$$

Where:

$(\Delta NO_3)_{AN}$ = Change in nitrate-nitrogen concentration during anaerobic period, mg/L

$(\Delta NO_2)_{AN}$ = Change in nitrite-nitrogen concentration during anaerobic period, mg/L

V = Reactor volume, L

$$OC_{NIT} = [4.57\,(\Delta NO_3)_A + 3.43\,(\Delta NO_2)_A]V \qquad 10.3\text{-}3$$

Where:

$(\Delta NO_3)_A$	=	Change in nitrate-nitrogen concentration during aerobic period, mg/L
$(\Delta NO_2)_A$	=	Change in nitrite-nitrogen concentration during aerobic period, mg/L

$$OCR = \frac{OC_{CARB}}{QS_O - QS} \qquad 10.3\text{-}4$$

Where:

OCR	=	Oxygen consumption ratio, g O_2 consumed for carbonaceous removal per g COD applied, g/g
Q	=	Batch feed volume per cycle, L
S_O	=	Influent total COD concentrating mg/L
S	=	Effluent soluble COD concentration, mg/L

The OCRs for R1 and R2 are compared in Table 10-12 for four OUR tests done at the end of July. OUR tests were done on both R1 and R2 using the same primary effluent feed plus their respective supplemental COD. The results in Table 10-12 show no evidence of COD loss for the lactate-fed reactor (R1) within the variability of the test results, with OCR values averaging 0.45 mg O_2/mg COD for both R1 and R2. Excluding the July 29 value with the much higher OCR, the average OCR value for R2 would be decidedly lower than that for R1 (0.39 mg/mg), which is an opposite finding to that expected if a COD loss was occurring in R1. Without the July 29 data for R2, the comparative OCRs for R1 and R2 range from 0.43–0.49 and 0.36–0.42, respectively. The range for R2 is below R1, indicating no clear evidence of COD loss for R1.

The variability for the OUR data is significant considering that over half the feed COD was from the supplemental COD, which was added on a relatively uniform basis. For R1 the range of OCRs varied by 5–12% and for R2 by 20–40% including the July 29th data. Without the July 29th data, the variation in OCR for R2 is about 8%.

Table 10-12. Comparison of OCRs Determined from OUR Tests for R1 (Lactate) and R2 (Acetate/Prop) Anaerobic/Aerobic SBR Operation.

DATE	OC_{OUR}, mg	OC_{DN}, mg	OC_{NIT}, mg	OC_{CARB}, mg	CODr, mg	OCR, mg/mg
R1, Lactate						
July 25	992	4	151	843	1732	0.49
July 29	913	4	195	722	1600	0.45
July 30	978	1	180	797	1872	0.43
July 31	981	0	185	796	1808	0.44
R1 Average ± SD						0.45 ± 0.03
R2,Acetate/Pro						
July 25		1	176	681	1632	0.42
July 29		3	173	956	1524	0.63
July 30		0	185	632	1780	0.36
July 31		0	212	710	1808	0.39
R2 Average ± SD						0.45 ± 0.12
R2 Average ± SD, without July 29						0.39 ± 0.03

Similar to previous reports on COD loss, a COD balance was done with SBR OUR data. As in the method presented by Randall et al (1992) the COD removal in the system should equal the COD measured by the carbonaceous oxygen uptake and the COD in the solids wasted from the systems if no COD loss exist:

$$COD_r = OC_{CARB} + COD_{WS} \qquad 10.3\text{-}5$$

$$COD_{WS} = F_{COD}\,(P_{VSS}) \qquad 10.3\text{-}6$$

Where:

COD_r = COD removal, mg/d

COD_{WS} = COD in waste solids, mg/d

F_{COD} = 1.4 g COD/g VSS

P_{VSS} = Waste solids per cycle, mg VSS

If no COD loss exists, the COD_r calculated from Equation 10.3-5 would equal the COD removal based on the COD added to the SBR minus the effluent soluble COD. A number of assumptions were necessary to carry out this calculation. First the COD content of the waste solids and the amount of the waste solids, to use in the COD balance with the SBR OC_{CARB} data had to be determined. COD analyses were done on the mixed liquor during the OUR tests and are summarized in Table 10-13.

Table 10-13. Summary of COD Analyses of Mixed Liquor During OUR Tests.

Date	MLVSS mg/L	TCOD mg/L	COD/VSS mg/L
R1, Lactate			
July 25	1710	2380	1.39
July 29	1630	2448	1.40
July 30	1670	2288	1.37
July 31	1670	2368	1.42
			Average 1.40
R2, Acetate/Prop			
July 25	1580	2166	1.37
July 29	1630	2202	1.35
July 30	1560	2122	1.36
July 31	1570	2142	1.36
			Average 1.36

Based on these results a COD/VSS ratio (F_{COD}) of 1.4 was used for the COD mass balance calculation.

The assumption for the amount of waste solids in the COD balance calculation was based on the SBR average SRT and the assumption that equal amounts of solids were produced per each SBR cycle. Thus the value for P_{vss} was estimated as follows:

$$P_{VSS} = \frac{(V)(X\ )}{(SRT)Nc} \qquad 10.3\text{-}7$$

Where:

X = SBR MLVSS concentration at full volume, mg/L

Nc = Number of cycles per day (5 cycles/day).

With these assumptions and data collected for R1 and R2 SBRs, the COD mass balance results are summarized in Table 10-14.

Table 10-14. Comparison of COD Mass Balance from Oxygen Uptake Rate and Solids Production Data to Measure COD Removal for R1 and R2 SBRs.

Date	Measured COD_r, mg	COD_{WS}, mg	OC_{CARB}, mg	Calculated COD_r, mg	COD Recovery, %
R1					
July 25	1732	885	843	1728	100
July 29	1600	957	722	1679	105
July 30	1872	913	797	1710	91
July 31	1808	935	796	1731	96
Average ± SD					98 ± 6
R2					
July 25	1632	885	881	1566	96
July 29	1524	913	956	1869	123
July 30	1780	874	632	1506	85
July 31	1808	879	710	1589	88
Average ± SD					98 ± 17

The percent COD recovery is based on the ratio of the calculated COD removal using the OUR data to the measured COD removal based on the total influent COD minus the effluent soluble COD concentrations and feed volume of 4L. For the R1 SBR, supplemented with lactate feed, the COD recovery ranged from 91–100% with a correlation coefficient of about 6%. For the acetate-fed reactor the COD recovery ranged from 85–123% with a correlation coefficient of about 16%. However, as was shown in Table 10-12, the July 29 date for R2 appears to be an outlier. Without including that date, the average COD recovery for R2 is 90%, ranging from 85%–96%, with a correlation coefficient of 9%. These results, using the same COD mass balance approach of previous investigators, do not provide evidence for a COD loss for the A/O process.

In August the operation was modified to compare the solids production for an A/O system to that for an aerobic system only. These tests used the same feed as during the first phase of testing at West Point, which consisted of primary effluent and supplemental COD from lactate. On August 12, the reactor contents of both systems were combined and equally distributed to each reactor. The reactors continued to be operated with the 4-day SRT. Table 10-15 summarizes the average reactor operating conditions for this phase. A lower effluent phosphorus concentration is shown for the A/O system compared to the aerobic-only system, indicating that biological phosphorus removal was achieved in the A/O system. However numerous operating problems prevented the maintenance of stable long-term operating conditions needed for a valid comparison of solids production between the two systems. In fact, the aerobic-only system showed a lower MLVSS concentration and lower solids production, but this was more likely related to the inability to accurately account for a high level of solids lost in the SBR effluent due to poor settling characteristics. Without the anaerobic contact phase to serve as a selector mechanism to control filamentous bacteria growth, the aerobic-only system became dominated by filamentous bacteria, and as a result had poor solids thickening properties and a higher sludge blanket level after the settling period. Solids often were pulled into the decant flow and a review of the data suggests that these solids were not always accounted for, possibly due to sampling errors. In addition to this problem the reactor feeding was interrupted for three days from August 31 to September 2 and two days from September 9 to September 10, due to a failure of the primary effluent feed pump that provided flow through reservoir from which the SBR feed pumps received the primary effluent. Also during this period nitrification toxicity was encountered, as evidenced by high effluent NH_4-N concentrations from both reactors.

Table 10-15. Average SBR Influent and Effluent Characteristics for August 26-September 11, 2002 during West Point Tests. Both Reactors Received Lactate as Supplemental COD. Reactor 1 is an Anaerobic/Aerobic System and Reactor 2 is an Aerobic System Only.

Parameter	Influent Reactor 1	Influent Reactor 2	Effluent Reactor 1	Effluent Reactor 2
Flow*, L/d	14.1 ± 9.4	14.7 ± 9.8		
PE TCOD, mg/L	303 ± 31	303 ± 31		
TCOD with supplement, mg/L	474 ± 33	469 ± 30	61 ± 28	120 ± 61
TSS, mg/L	73 ± 8	73 ± 8	25 ± 21	69 ± 36
VSS, mg/L	63 ± 10	63 ± 10	20 ± 17	60 ± 31
NH_4-N, mg/L	29 ± 4	29 ± 4	5.2 ± 7.2	4.7 ± 5.5
NO_3-N, mg/L			5.1 ± 0.7	4.4 ± 0.7
Sol PO_4-P, mg/L	15.6 ± 0.5	15.5 ± 0.4	1.0 ± 0.9	8.9 ± 1.3
Total P, mg/L	16.8 ± 1.9	16.8 ± 1.6	2.4 ± 1.6	10.1 ± 0.3
pH			7.3 ± 0.1	7.3 ± 0.1
Average Temperature, ^oC			18.0	18.0

* normally at about 20 L/day, but feed pump was off for three days

Based on the difficulties with the feed operation at West Point and in achieving a stable operation with the aerobic-only system, it was decided to continue the work on comparing solids production for the A/O and aerobic-only processes at the University of Washington laboratories, where closer monitoring and reactor control could be done. West Point primary effluent was delivered daily for reactor feeding and the feed was also supplemented with lactate. A major difference in the operation, as described in the Methods section, was that the reactors were fed once per day to simplify the operation and to provide maximum control and accountability of the feed characteristics.

10.3.2 Test #2 Results—Anaerobic/Aerobic versus Aerobic Treatment Only

In late September 2002, the bench-scale SBR units for Test #2 were seeded with an activated sludge containing a biological phosphorus-removing bacteria population. Both reactors were batch-fed once per day with West Point primary effluent supplemented with 272 mg/L COD as lactate. The total COD of the primary effluent averaged 306 mg/L (Table 10-16), so the lactate feed represented over half of the feed biodegradable COD. The A/O system was operated as in the West Point tests with nitrogen sparging in the anaerobic period to assure anaerobic conditions. The ORP during the anaerobic period was monitored at value in the range of -100 to -200 mV, indicating anaerobic conditions were maintained.

Table 10-16. Test #2 Average SBR Influent and Effluent Characteristics for Anaerobic/Aerobic (A/O) and Aerobic (O) Reactors Operated with Once per day Feeding at 10-day SRT with Lactate Addition (272 mg/L). October 23 to November 11, 2002 after 2SRT Operating Period.

Parameter	Influent	Effluent A/O Reactor	Effluent O Reactor
Flow, L/d	3.0		
PE TCOD, mg/L	306 ± 39		
TCOD with supplement, mg/L	578 ± 39		
TSS, mg/L	89 ± 25	32 ± 12	31 ± 5
VSS, mg/L	68 ± 16	25 ± 9	25 ± 4
pH		7.8 ± 0.1	7.8 ± 0.1
Average Temperature, ^{0}C		20.0	20.0

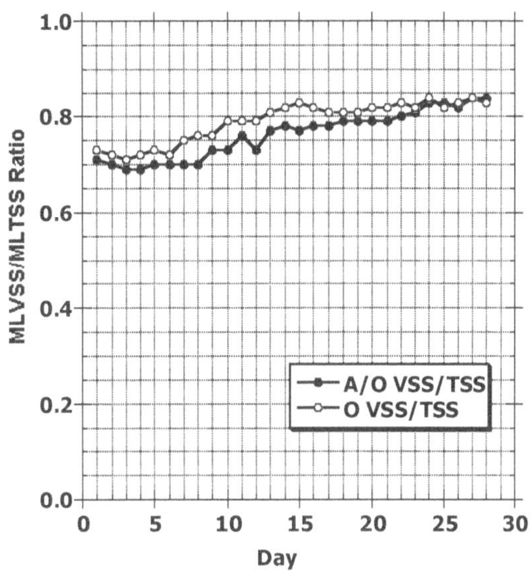

Figure 10-8. MLVSS/MLSS Ratio Versus Time for Laboratory Tests with Anaerobic/Aerobic (A/O) and Aerobic (O) Systems Operated With Once Per Day Feeding and 10-day SRT.

In spite of the anaerobic conditions for the first hour of the A/O operation, biological phosphorus removal could not be maintained. After seeding, the effluent soluble phosphorus concentration steadily increased to match that of the aerobic only SBR (12–17 mg/L) process within two weeks, and orthophosphorus release in the anaerobic period was only 4–8 mg/L. Even though the biological phosphorus removal mechanism was lost, soluble COD removal still occurred at 80–90% in the anaerobic period. The loss of PAO activity and growth with time is illustrated by the increased MLVSS/MLSS ratio in Figure 10-8. As was shown in the West Point tests, systems with appreciable biological phosphorus removal have a lower MLVSS/MLSS ratio due to the storage of cations with the polyphosphates. For Test #2 the VSS/TSS ratio of the seed mixed liquor was at 0.68–0.70 g/g. After 21 days, the ratio increased to 0.75 and finally to near 0.85 and equal to the aerobic system VSS/TSS ratio. The rate of increase in the VSS/TSS ratio was faster for the aerobic system, indicating that some PAO growth was occurring, though at a steadily diminishing rate, in the A/O system.

Microscopic examination of mixed liquor samples from the A/O and O systems at 1000-magnification under phase contrast revealed a high population of a tetrad coccoid cell formation in the A/O system but not in the O system. The tetrad formation is characteristic of G bacteria (Cech and Hartman, 1990), which are able to out-compete PAOs in some cases for substrate uptake in the anaerobic contact period of EBPR systems and thus prevent PAO growth. Substantial soluble COD removal was observed in the anaerobic period, with little if any biological phosphorus removal, which agrees with the microscopic observations on G bacteria. Examination of literature on G bacteria suggests that the once-per-day feeding and longer SRT may have favored growth by G bacteria. Liu et al. (1997) claimed that excessive aeration times depleted glycogen storage in PAOs, making them less competitive with G bacteria. It should be noted that the presence of G bacteria does not preclude the fact that fermentation conditions still existed in the anaerobic contact period. The difference from an EBPR system is that the substrate was taken up by G bacteria instead of PAOs.

After a time period equal to 2 SRTs (20 days), the solids production and observed solids yield was evaluated for the two parallel reactors over a 20-day period from October 23 to November 11. The reactors exhibited similar performance in effluent TSS/VSS and pH as summarized in Table 10-16. Both reactors were maintained close to their target SRT of 10 days and they both had similar average MLVSS (866 and 845 mg/L) and MLTSS (1083 and 1035 mg/L) concentrations, and VSS/TSS ratios (0.80 to 0.81 g/g), as summarized in Table 10-17. The daily measured MLVSS concentrations are very similar, and in most cases within 5% (Figure 10-9).

Table 10-17. Average Solids Yield Values for SBRs for Test #2 Tests from October 23 to November 11, 2002 after 2SRT Operating Period. Anaerobic/Aerobic (A/O) and Aerobic (O) Reactors Operated with Once Per Day Feeding at 10-day SRT with Lactate Addition (272 mg/L).

Parameter	A/O Reactor	O Reactor
Daily COD removal*, mg/d	1644 ± 117	1644 ± 117
MLVSS, mg/L	866 ± 87	845 ± 103
MLSS, mg/L	1083 ± 65	1035 ± 109
MLVSS/MLSS	0.80 ± 0.04	0.81 ± 0.02
SRT, d	10.0 ± 0.24	10.0 ± 0.22
Daily VSS production, mg/d	475 ± 151	491 ± 135
Yield, g VSS/g CODr	0.28 ± 0.08	0.29 ± 0.07

*Influent TCOD minus effluent soluble COD

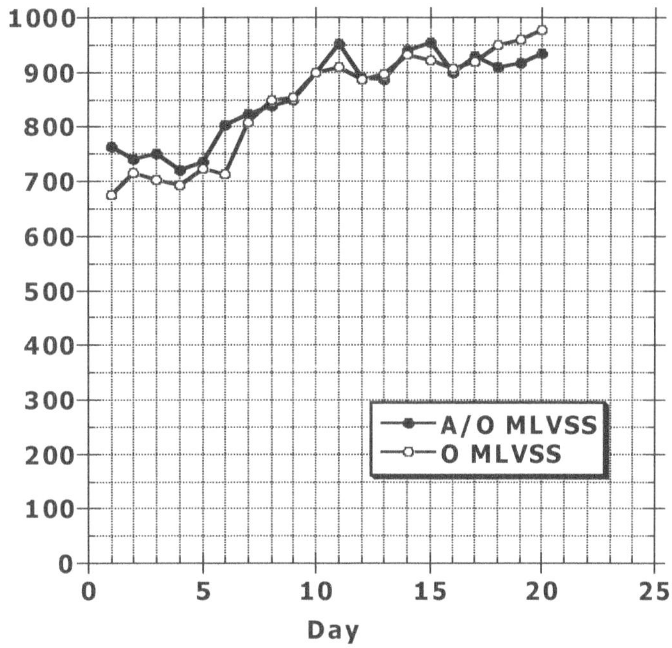

Figure 10-9. Comparison of MLVSS Concentrations for Anaerobic/Aerobic (A/O) and Aerobic (O) Reactors Treating Primary Effluent Plus Lactate With 10-day SRT After Operating Time Equal to 2SRTs.

The average solids production was calculated daily from the reactor data and included changes in the reactor solids inventory [(MLVSS day, $_n$ – MLVSS day, $_{n-1}$) (Volume)] plus the solids removed by manual wasting and in the effluent. The average observed yields shown in Table 10-17 for the 20-day period are very close at 0.28 and 0.29 g VSS/g COD for the A/O and O reactors, respectively. The daily yield values follow similar trends with no one reactor yield value consistently higher or lower than the other (Figure 10-10). The variations in the yields likely are due to changes in the primary effluent feed nonbiodegradable VSS concentrations and experimental variations in sampling and measurement.

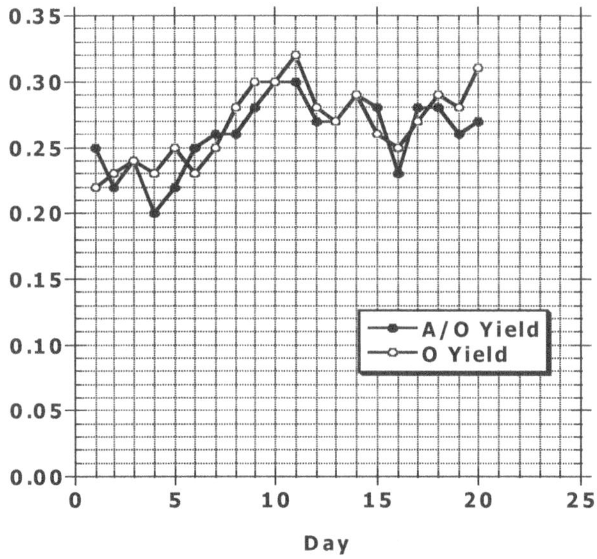

Figure 10-10. Comparison of 5-day Running Average of Observed Yields for Anaerobic/Aerobic (A/O) and Aerobic (O) Reactors Treating Primary Effluent Plus Lactate With 10-day SRT After Operating Time Equal to 2SRTs.

10.3.3 Test #3—Anaerobic/Aerobic Versus Aerobic-Only Treatment with Synthetic Wastewater

The results for the laboratory tests with a synthetic feed are divided into three parts. The first (Test #3A) is for testing with only acetate as the COD source for both the A/O and O systems. The next two tests (Test #3B and #3C) are with different proportions of lactate in the feed. Results of a steady-state operating condition for the two processes during May 2003 for Test #3A are presented first. After May 27, the feed was modified to 50% acetate and 50% lactate. The results from this testing condition are referred to as Test #3B and the intent was to determine if fermentation of lactate in the anaerobic zone may have affected COD loss in the A/O system to result in a lower VSS production compared to the O system without anaerobic fermentation. From June 23 to July 5, the parallel operation continued with the feed modified to 100% lactate (Test #3C), so that COD loss due to fermentation might be more easily measured.

In April 2003, stable operating conditions were developed for the laboratory anaerobic/aerobic (A/O) reactor treating synthetic wastewater with 100% acetate as the main COD source. ORP values during the anaerobic contact period were in the -100 to -150 mV range, about 90% of the soluble COD was consumed during anaerobic contacting, and orthophosphorus release was very high, in the range of 60–70 mg/L as P versus an influent P concentration of about 12 mg/L. At the end of April the aerobic-only system was started using waste sludge from the A/O system. The target SRT for both systems was six days but the actual SRT was closer to 5.5 days, based on the waste sludge calculation that included sludge wasted during sampling as well as effluent suspended solids and daily manual sludge wasting.

Following operation for a time period equal to 2 SRTs, results were obtained for a 15-day quasi- steady-state period in May to compare the reactor performance and VSS production for the A/O and O systems with only acetate as the COD source (Tables 10-18 and 10-19). The feed to both reactors was from a common reservoir so that they both had identical feed characteristics. The feed volume was monitored closely and maintained at 1.0 L per cycle. The consistent high level of biological phosphorus removal is demonstrated for the A/O system with an average effluent P concentration of 0.30 mg/L and average P concentration of 65 mg/L in the anaerobic contact period (Table 10-18). The average effluent P concentration of 10.6 mg/L for the O system versus an influent concentration of 11.7 mg/L is representative of phosphorus removal for only cell synthesis. The data show that the reactors had similar average effluent VSS concentrations and a slightly higher average TSS concentration for the A/O system.

Table 10-18. Test #3A Average SBR Influent and Effluent Characteristics for Anaerobic/Aerobic (A/O) and Aerobic (O) Reactors Operated with Five Cycles Per Day Feeding with Synthetic Wastewater at 100% Acetate. May 12 to May 27, 2003 After 2SRT Operating Period.

Parameter	Influent	Effluent A/O Reactor	Effluent O Reactor
Flow, L/d	5.0		
Soluble COD, mg/L	188		
Percent Acetate, %	100		
Percent Lactate, %	0		
Orthophosphorus, mg/L	11.7	0.3 ± 0.3	10.6 ± 1.8
Anaerobic period orthophosphorus, mg/L		65 ± 10	N/A
TSS, mg/L		37.6 ± 6.2	33.1 ± 7.4
VSS, mg/L		25.6 ± 3.6	26.7 ± 5.1
SRT, d		5.6 ± 0.27	5.4 ± 0.44
pH		7.6 ± 0.1	7.6 ± 0.1
Average Temperature, ^{0}C		15.0	15.0

N/A, not applicable

With regard to biomass production, the results shown in Table 10-19 suggest a similar observed VSS yield and almost identical MLVSS concentrations for the A/O and O systems. The MLVSS concentrations average 971 and 986 mg/L for the A/O and O systems, respectively, a difference of only 1.5%. The yields were almost identical at 0.35 and 0.37 g VSS/g COD. The slightly higher yield for the O reactor may be related to the lower average SRT of 5.4 versus 5.6 days, but most likely it reflects experimental variability. The A/O system had a much higher MLSS concentration and lower mixed liquor VSS/TSS ratio (0.61 versus 0.83 g/g). The lower VSS/TSS ratio is due to the storage of Ca^{++}, Mg^{++}, and K^+ cations along with the phosphorus uptake and polyphosphate storage in the EBPR process.

Table 10-19. Average Solids Yield Values for SBRs for Test #3A from May 12 to May 27, 2003 after 2SRT Operating Period. Anaerobic/Aerobic (A/O) and Aerobic (O) Reactors Operated with Five Cycles Per Day with Synthetic Wastewater at 100% Acetate.

Parameter	A/O Reactor	O Reactor
MLVSS, mg/L	971 ± 86	986 ± 103
MLSS, mg/L	1590 ± 128	1192 ± 107
MLVSS/MLSS	0.61 ± 0.03	0.83 ± 0.07
SRT, d	5.6 ± 0.27	5.4 ± 0.44
Yield, g VSS/g CODr	0.35 ± 0.03	0.37 ± 0.04

After data collection for the acetate feed COD experiment ended on May 27, the feed was modified to 50% acetate and 50% lactate (Test #3B). The total feed soluble COD concentration was kept the same at 188 mg/L. Between May 27 and June 10 a transition period occurred with the SRT increased to eight days. A reactor upset also occurred during this period in which solids overflowed due to the failure of a decant pump. Following the solids overflow on June 8, the solids were returned to the reactors and solids from the A/O reactor were added to the O reactor in an attempt to start at an equal VSS concentration (940 mg/L for the O system versus 1060 mg/L for the A/O system). At the end of the period the A/O system still had a 14% higher MLVSS concentration, at about 1400 mg/L versus 1220 mg/L for the O system. The average operating characteristics for the two systems for Test #3B from June 10 and June 23 are summarized in Table 10-20. As shown by the effluent P and anaerobic contact period P concentrations, the change to an acetate/lactate feed did not have any noticeable effect on the biological phosphorus removal system performance. The effluent P concentration averaged 0.30 mg/L, which is close to that for Test #3A with 100% acetate COD at the same influent concentration. The P release concentration was also similar. Again, both reactors had similar effluent VSS concentrations and similar pH values.

Table 10-20. Test #3B Average SBR Influent and Effluent Characteristics for Anaerobic/Aerobic (A/O) and Aerobic (O) Reactors Operated with Five Cycles Per Day Feeding with Synthetic Wastewater at 50% Acetate and 50% Lactate. June 10 to June 23, 2003 After One SRT Operating Period. SRT Maintained at 8.0 Days.

Parameter	Influent	Effluent A/O Reactor	Effluent O Reactor
Flow, L/d	5.0		
Soluble COD, mg/L	188		
Percent Acetate, %	50		
Percent Lactate, %	50		
Orthophosphorus, mg/L	11.7	0.30 ± 0.5	11.3 ± 0.4
Anaerobic period orthophosphorus, mg/L		60 ± 8	N/A
TSS, mg/L		24.2 ± 5.3	23.5 ± 3.3
VSS, mg/L		20.1 ± 4.2	21.5 ± 3.2
SRT, days		8.0 ± 0	8.0 ± 0
pH		7.6 ± 0.1	7.6 ± 0.1
Average Temperature, ^{0}C		15.0	15.0

N/A, not applicable

If the COD loss mechanism is real and related to fermentation, the addition of lactate in Test #3B should have resulted in a lower observed VSS yield coefficient for the A/O reactor versus the O reactor. However, the results (Table 10-21) show a higher average observed yield (14% higher) and higher average MLVSS concentration (16% higher) for the A/O reactor versus the O reactor. The average observed yields were 0.31 and 0.27 g VSS/g COD and the average MLVSS concentrations were 1222 and 1052 mg/L, for the A/O and O systems, respectively.

During Test #3C with 100% lactate as the COD source, the biological phosphorus removal efficiency for the A/O system remained high with an average effluent P concentration of 0.10 mg/L during the period June 24–July 5, 2003. The solids production results for this period are summarized in Table 10-22. As in Test #3B, the biomass yield value was higher for the A/O system and its average MLVSS concentration was higher, by about 20%, compared to the O system. The biomass yields are 0.36 and 0.30 g VSS/g COD, for the A/O and O systems, respectively. The average MLVSS concentrations are 1449 and 1194 mg/L, with standard deviations of 58 and 73 mg/L, respectively, suggesting that the difference between the A/O and O systems is significant.

Table 10-21. Average Solids Yield Values for SBRs for Test #3B from June 10 to June 23, 2003 After One SRT Operating Period. Anaerobic/Aerobic (A/O) and Aerobic (O) Reactors Operated with Five Cycles Per Day with Synthetic Wastewater at 50% Acetate and 50% Lactate.

Parameter	A/O Reactor	O Reactor
MLVSS, mg/L	1222 ± 135	1052 ± 124
MLSS, mg/L	1845 ± 283	1198 ± 140
MLVSS/MLSS	0.67 ± 0.05	0.88 ± 0.01
SRT, days	8.0 ± 0	8.0 ± 0
Yield, g VSS/g CODr	0.31 ± 0.03	0.27 ± 0.03

Table 10-22. Average Solids Yield Values for SBRs for Test #3C from June 23 to July 5, 2003. Anaerobic/Aerobic (A/O) and Aerobic (O) Reactors Operated with Five Cycles Per Day with Synthetic Wastewater at 100% Lactate.

Parameter	A/O Reactor	O Reactor
MLVSS, mg/L	1449 ± 58	1194 ± 73
MLSS, mg/L	2294 ± 80	1379 ± 89
MLVSS/MLSS	0.63 ± 0.02	0.87 ± 0.02
SRT, days	8.0 ± 0	8.0 ± 0
Yield, g VSS/g CODr	0.36 ± 0.01	0.30 ± 0.02

10.4 Discussion

A number of experiments were done to assess the presence of COD loss in an A/O process. In contrast to most of the previous work that formed the basis for COD loss in A/O processes, two reactors were operated in parallel to allow testing a different condition suspected to be related to COD loss. Previous work relied on a COD balance from a single reactor for the determination of a COD loss. The observed yield values for the various experiments in this research are summarized in Table 10-23 and is no case is there a reduction in solids yield to support the existence of a COD loss.

For the West Point tests the primary effluent feed to one of the A/O reactors was supplemented with acetate, a non-fermentable substrate, and the other A/O reactor received lactate. Assuming that the COD loss for A/O systems is the result of a fermentation product, such as hydrogen escaping in the fermentation process, a lower yield value was expected for the system fed only lactate. However, for both sets of data the average observed yields between the systems were almost identical. One issue is whether the supplemental COD was added at a high enough concentration to allow a measurable difference in solids production.

The supplemental COD addition during the West Point tests was significant at 173–197 mg/L, but the amount of acetate added may not have been great enough to provide a sufficient difference in biomass production, assuming that COD loss is related to fermentation. For example, for the July testing the average VSS production normalized to the influent flow and based on the observed solids yield and COD removal was 193 mg/L. Comparing the two reactors in the West Point tests, the amount of COD that was subject to a COD loss (acetate versus lactate addition) in one reactor and not in the other due to fermentation was 131 mg/L. An approximate yield of 0.30 g VSS/g COD can be applied to synthetic substrates based on results from Test #3 shown in Table 10-23. Assuming a 25% COD loss in the anaerobic contact period, which is within the median range of the COD loss reported in the previous work, the solids production could be decreased by about 9.8 mg/L (131 x 0.30 x 0.25), which is only 5% of the total VSS production. Considering that the standard deviations of the yields in Tables 10-9 and 10-10 were 11–14% of the mean, the amount of supplemental claimed COD was too low to provide any conclusions about COD loss based on VSS measurements for the West Point tests.

Table 10-23. Summary of Average Observed Solid Yield Values for Different Test Conditions Used to Evaluate COD Loss.

Tests	Observed Solid Yield, g VSS/g COD_r	
	A/O	Aerobic-Only
WP Test – June		
PE + Lactate	0.37 ± 0.05	-
PE + Acetate/Prop	0.37 ± 0.05	-
WP Test – July		
PE + Lactate	0.44 ± 0.05	-
PE + Acetate/Prop	0.43 ± 0.05	-
Test #2		
PE + Lactate	0.28 ± 0.08	0.29 ± 0.07
Test #3A		
Acetate	0.35 ± 0.03	0.32 ± 0.04
Test #3B		
50% Acetate, 50% lactate	0.31 ± 0.03	0.27 ± 0.03
Test #3C		
100% Lactate	0.36 ± .01	0.30 ± .02

However, the COD balance results, which were done using the same approach as in the previous work reported in the literature, showed no COD loss for the West Point tests. As in previous work, COD balances were done using the OUR test data, solids wasting data, and influent and effluent COD data. As was shown in Table 10-14, which summarized the COD balance calculations for four sets of OUR data, the COD balances averaged close to 100% recovery, indicating no COD loss, in spite of a substantial amount of fermentable COD fed to the lactate reactor. The COD balance recoveries were similar for both the lactate and acetate fed reactors, suggesting no COD loss for the A/O system regardless of the source of acetate for EBPR.

The inability to observe COD loss for the A/O process as reported for other A/O systems was perplexing. One explanation for the lack of observed COD loss in the current experiments may be the use of nitrogen sparging during the anaerobic contact period in the reactors. Nitrogen sparging would prevent oxygen transfer from the gas space above the anaerobic contact reactor liquid. Any oxygen transferred by this pathway and used in the anaerobic contact zone or period would not be accounted for by the COD balance calculation procedure, which relied only on oxygen uptake in the aerobic zones. Relatively shallow reactors used for laboratory or pilot plant tests have a high air/liquid surface interfacial area to volume ratio in comparison to full-scale systems, and thus offer a significant oxygen transfer potential. In previous work using laboratory reactors to study COD removal in EBPR processes, nitrogen sparging was used in the anaerobic contact period to obtain COD/P ratios for EBPR that was mainly due to COD uptake by the PAOs (Stephens and Stensel, 1998). Previous studies on COD loss (Randall et al., 1992; Wable and Randall, 1994; Barker and Dold, 1995.) have not provided information on possible transfer of oxygen into the anaerobic contact zone. ORP measurements, for example, would have been helpful to support the fact that substantial anaerobic conditions existed. In the current work, ORP measurements were used to confirm that proper anaerobic conditions existed. During the early phases of our study, the project team found that EBPR could not be achieved until oxygen transfer during the anaerobic contact period was prevented. In systems without strict control of oxygen transfer into the anaerobic zone, a sufficiently high influent COD/P ratio can provide EBPR in spite of oxygen transfer during the anaerobic period. In this way some COD is oxidized using the oxygen transferal, but there is still sufficient COD for high phosphorus removal by PAOs. In such cases, a zero DO concentration can be measured in spite of oxygen transfer and utilization. Since OURs were only evaluated in the aerobic zones in the previous work, the amount of oxygen used—and thus COD oxidized in the anaerobic zones—was not ascertained by any measurements.

Though oxygen transfer into the anaerobic contact zone may help explain COD loss calculations in previous work, it does not explain why Barker and Dold (1995) found COD loss for tests with bench-scale EBPR systems treating wastewater but not for such systems receiving acetate as the carbon source. However, details of reactor design and operation were not given and the data for these comparisons were not obtained from studies with parallel reactor operations. Thus, without more information, it is difficult to determine if the COD loss was related to the reactor design and operation or to the type of feed COD.

Solids production for A/O and O systems operated in parallel were compared in Tests #2 for a feed that was similar to that used for the West Point tests, consisting of primary effluent and lactate addition. The average observed yield values were almost identical for the A/O and O systems and lower than that for the West Point Test. The difference may have been due to using a longer SRT in Test #2 (10 days versus four days) and to changes in the primary effluent (PE) feed due to the feed collection and time of year. For Test #2 the PE was obtained on a grab basis versus feeding from the primary effluent over a 24-hour period during the West Point Tests. Based on the COD added and observed yield the average VSS production was about 153 mg/L. Assuming a 25% COD loss for the lactate addition (272 mg/L COD) and approximately 90 mg/L of influent rbCOD from the domestic wastewater and a net yield of 0.30 g VSS/g COD, the VSS production should have been lower by about 27 mg/L or 18%. Considering that the MLVSS data shows a standard deviation of about 10–12%, the expected COD loss should have been measured by a difference in the average MLVSS concentrations for the A/O and O systems and average observed yields. Thus no significant COD loss can be concluded for Test #2. The fact that the A/O system in this test was dominated by a non-PAO population does not change the significance of these results, as all other anaerobic reactions normally occurring in the anaerobic period would have occurred.

No COD loss was expected for Test #3A as acetate was the only carbon source and is not fermentable. The observed yields for the A/O and O systems averaged 0.35 and 0.32 g VSS/g

COD, respectively. The yield was about 10% higher for the A/O system but this is not significant considering the accuracy of the VSS test. An important feature of Test #2 is that it demonstrates the ability to measure similar biomass yields for the two parallel A/O and O reactors under conditions with no COD loss and similar VSS yields expected. These results verify the laboratory techniques and give credit to other tests where solids yields are compared.

No COD loss can be claimed for Tests 3B and 3C, as the average yield at 50% and 100% lactate, respectively, was higher for the A/O system. A lower yield would be expected for the A/O system for Tests 3B and 3C if an anaerobic COD loss were occurring due to hydrogen loss in fermentation, as lactate is fermentable and acetate is not. The higher yield for the A/O system is difficult to explain, but may be related to carbon storage products for the A/O system biomass versus the O system biomass, and the longer SRT used in Tests 3B and 3C versus Tests 3A (8 days versus about 5.5 days). In the A/O process the COD is first assimilated and converted to intracellular storage products (such as polyhydroxybutyrate, PHB) by the PAOs, which is oxidized in the aerobic period. Bordacs and Tracy (1988) hypothesized that the COD of these storage products is metabolized more slowly under aerobic conditions than if the initial substrate was consumed directly under aerobic conditions. Thus it is conceivable that at the longer SRT the effect of the storage products on the solids mass may have been more noticeable than at the lower SRT.

The results for Tests 3B and 3C also suggest that the fraction of lactate affected the solids yield in the A/O system. Perhaps lactate caused the formation of a different storage product that was not as easily metabolized compared to the carbon storage products produced from acetate assimilation. For example, lactate can be fermented to propionate, which can be assimilated, by PAOs and possibly result in a different carbon storage product. In addition, some storage products could have been formed in the O system due to the rapid batch feeding of the COD at each SBR cycle. Analysis of the biomass composition and storage products for each reactor system would have been helpful, but the additional resources needed for such work was not available for this study.

The COD loss mechanism is further reviewed here to determine whether any unique features of our experimental design prevented its occurrence and subsequent lower biomass production. Because COD involves equivalent electron transfer for the oxidation of any substance, the feed COD to a biological process must be accounted for by oxidation or by the final products: COD in the effluent liquid, solids removed, and gas emissions. Besides aerobic oxidation of substrate, COD removal by oxidation can occur with other electron acceptors, such as sulfate, nitrate/nitrite, and CO_2 with CH_4 formation. Nitrate and nitrite have been accounted for in the COD balance calculations, and results in this work showed that sulfate reduction is not occurring in A/O systems. The discussion in Section 4.2 points out that methane production is not likely due to the short anaerobic SRT expected in A/O processes, even if the methanogenic organisms could survive the effects of oxygen exposure. To account for COD loss then, what remains is either measurement error(s) or the loss of COD due to the transformation of substances to volatile compounds and transfer to the gas phase. The most likely volatile compound to be produced in the anaerobic step that could be stripped out is hydrogen. The production of volatile fatty acids is not considered a source of COD loss due to their consumption by PAOs and their high solubility and relatively low Henry's coefficient. Hydrogen production and release is evaluated further here, especially with regard to its potential formation from lactate, which was used as a synthetic substrate in this study, and it formation from domestic wastewater compounds.

A wide variety of organisms can ferment lactate, and the fermentation products depend on the organisms involved. Zellner et al. (1994) has reported on the types of lactate fermentation reactions carried out by specific organisms, and these are illustrated in Table 10-24.

Table 10-24. Fermentation Reactions with Lactate (Zellner, 1994).

Reaction	Microorganism
(1) 2 Lactate → 3 Acetate	*Clostridium formicaceticum*
(2) Lactate + H_2O → Acetate + CO_2 + $2H_2$	*Desulfovibrio vulgaris*
(3) 3 Lactate → 2 Propionate + acetate + CO_2 + H_2O	*Clostridium propionicum* *Desulfobulbus propionicus*

Reaction 2 is the only one in Table 10-24 that produces hydrogen directly from lactate, with about 33% of the lactate COD converted to hydrogen. If this reaction was occurring in our study and the hydrogen was effectively stripped due to the nitrogen sparging, a significant COD loss would have occurred with a measurable reduction in biomass production compared to the aerobic only system; however, this result was not observed.

A more commonly expected pathway is the conversion of lactate to propionate and acetate (Ballapragada, 1996), as shown by Reaction (3) in Table 10-24. Hydrogen is then produced by propionate fermentation:

$$\text{Propionate} + 2H_2O \rightarrow \text{Acetate} + 3H_2 + CO_2 \qquad 10.4\text{–}1$$

The combination of Reaction (3) in Table 10-24 and the fermentation of propionate results in about 33% of the COD being converted to hydrogen. COD loss (and thus reduced sludge production) via this pathway requires that the propionate be fermented to acetate with more than 99% removal of hydrogen from the fermentation reactor, presumably by stripping. The lack of COD loss for the A/O system versus the O system in Tests 3B and 3C means that either hydrogen was not stripped or that propionate was consumed, which is very possible.

Propionate is a key fermentation intermediate that precedes hydrogen production. Propionate degradation to acetate, hydrogen, and carbon dioxide is thermodynamically favorable only when the hydrogen partial pressure is less than 10^{-4} atm, which is equal to an equilibrium liquid concentration of 75 nmol/L or 0.0015 mg/L (Zehnder, 1988). In anaerobic systems with methanogenesis, this low hydrogen concentration is maintained by interspecies hydrogen transfer, which is made possible by the close spatial distribution between fermenters and methanogens. This is termed a syntrophic relationship between fermenters and methanogens (McCarty and Smith, 1986) as it assure that the hydrogen fermentation can proceed with growth of both fermenters and methanogens. In anaerobic digesters used for solids processing, about 30% of the COD is converted to methane via hydrolysis, fermentation, and methanogenesis from hydrogen production as a result of propionate and butyrate production and fermentation (Tchobanoglous et al., 2003).

The fate of propionate COD in A/O systems is more likely to be propionate consumption by PAOs rather than fermentation to hydrogen and then COD loss due to hydrogen stripping. Though hydrogen is a poorly soluble gas with a relatively high Henry's constant, the ability to maintain the extremely low hydrogen concentrations of less than 0.001 mg/L needed for propionate fermentation is remote for the laboratory nitrogen-sparged reactor. It would be impossible to maintain such a low hydrogen concentration without methanogenesis in a full-scale wastewater treatment plant application with deep tanks and only mechanical unity to keep solids suspended. In addition, propionate can be assimilated by PAOs, and this would negate the possibility for hydrogen production. In a research paper on the evaluation of carbon storage products in PAOs, Liu et al. (2002) have shown that propionate can be used directly by the PAOs in addition to acetate.

Thus, the more likely pathway for the lactate fermentation was the production of propionate and acetate according to Reaction (3) in Table 10-24 and the direct consumption of the acetate and propionate by the PAOs. The ability to maintain the extremely low hydrogen concentration needed for propionate conversion to hydrogen is very unlikely. In addition, the PAOs would likely out-compete propionate fermenters, as they have a considerably higher biomass yield as aerobic growth organisms compared to the low biomass yields associated with anaerobic organisms.

Acetate and propionate are common fermentation products for organic compounds. Madigan et al. (1997) show both propionate and acetate production from the fermentation of amino acids and other organic compounds. Propionate and acetate are common fermentation products found at biological nutrient removal (BNR) facilities (Liu et al., 2002), with acetate more prevalent. At a full-scale Bardenpho facility in Orange County, Florida the ratio of acetate to propionate in the influent wastewater ranged from 2.0–5.4. Van Munch (1998) reported the volatile fatty acid composition for prefermenters at four full-scale BNR systems. The acetate content ranged from 49–71% and the propionate content ranged from 24–33%. Butyric and valeric acids accounted for the remainder. These data show that propionate was not being converted to acetate and hydrogen, and a likely explanation is that hydrogen was being maintained at a level such that propionate fermentation was unfavorable. The relative amounts of propionate and acetate are in the range of that expected for wastewater sludge fermentation to acetate and propionate and butyrate, and suggests that minimal amounts of propionate were being converted to hydrogen and acetate. This demonstrates that hydrogen stripping to low enough levels to encourage a flux of COD out of the system via propionate or butyrate fermentation was not practical in a full-scale reactor. The results also indicate that propionate is likely available in the anaerobic contact zone in the EBPR systems. However, it is very likely consumed by PAOs and not fermented further with hydrogen production and release from the system. These data for fermenters suggest that hydrogen was not removed sufficiently to allow fermentation of propionate to acetate.

Based on the results of this research and evaluation of fundamental mechanisms of substrate fermentation and COD utilization, there is no evidence to support COD loss in A/O processes commonly used for EBPR. The lack of COD loss and the inability to measure a lower VSS production was observed repeatedly in experiments with parallel A/O and O reactors. For parallel A/O and O systems fed with 100% lactate feed, the best case to observe COD loss was provided, but no reduction was found. The COD balance results for the West Point tests did not provide the same findings as previous studies, suggesting that some fundamental difference exists in the previous experimental set up or operation. However, the information on the test conditions used for the previous publications is not adequate to assess whether the COD loss observed was an artifact of the experimental conditions used.

The results of this work have important implications for the design of EBPR facilities. Using current activated sludge design models that incorporate COD loss could result in under-designed facilities, possibly underestimating oxygen requirements and solids production by 15–20%. Using a lower solids production estimate, engineers also may under-design the aeration rank volume. The current BioWin model includes a provision for hydrogen production via fermentation, but it does not address the need to have extremely low hydrogen concentrations in the reactor for this transformation pathway to be favorable. If this concept was included and a reaction pathway for propionate production and consumption by PAOs is also included, the model would not predict any COD loss.

10.5 Conclusions

Initial experiments were done with bench-scale sequencing batch reactors (SBRs) to test the hypothesis that COD loss in A/O systems is related to fermentation during anaerobic contact between the feed wastewater and mixed liquor. Two reactors were operated in parallel with anaerobic-aerobic sequences treating domestic primary effluent, with one spiked with lactate, a fermentable substrate, and the other with an equal amount of acetate. A lower observed solids yield for the acetate-spiked reactor was expected to indicate COD loss during fermentation. Following this experiment additional tests were done to evaluate solids production in Anaerobic/Aerobic (A/O) and Aerobic-only (O) systems operated in parallel treating primary effluent spiked with lactate and a synthetic wastewater at three conditions; 100% acetate, 50-50% acetate/lactate, and 100% lactate. The conclusions from the evaluation of these results are as follows.

- ☐ The observed solids yields in terms of g VSS/g COD were similar for the A/O systems operated in parallel at a 4-day SRT treating domestic wastewater spiked with lactate or acetate, indicating no effect from a greater level of fermentable substrate in the feed. No COD loss was found for these reactors using the same COD balance approach used by previous investigators reporting a COD loss for EBPR processes incorporating an A/O sequence.

- ☐ Sulfate reduction did not occur in the anaerobic period of the parallel A/O systems indicating the lack of sulfate reducing organisms, possibly due to oxygen inhibition in the aerobic period. Thus, sulfate reduction is not an important mechanism to explain COD loss in A/O systems.

- ☐ The observed solids yield was the same for an A/O and O system operated in parallel at a 10-day SRT treating primary effluent spiked with about 270 mg/L lactate, indicating no COD loss for the A/O system. In this case the A/O system did not accomplish biological phosphorus removal, but was instead dominated by G bacteria, which out competed PAOs and consumed soluble COD in the anaerobic contact period.

- The observed solids yields based on VSS for the A/O system were equal to or higher than that for the O system treating synthetic wastewater, regardless of whether it was 100% acetate, 100% lactate, or 50–50% acetate/lactate, to indicate no COD loss for the A/O system.

- Review of fundamental concepts on fermentation and biological phosphorus removal suggests that COD loss during anaerobic contact in an A/O process should not occur in full-scale systems via hydrogen production and stripping. Due to the relatively low gas/liquid mass transfer rates in such systems hydrogen can not be removed to maintain liquid concentrations at a level necessary to allow fermentation of propionate or butyrate to acetate and hydrogen. Thus, the production of propionate most likely is followed by propionate consumption by the PAOs.

CHAPTER 11.0

Summary and Conclusions

A wide range of processes that can reduce biomass production in the liquid treatment portion of biological wastewater treatment processes were identified and assessed. These processes included anaerobic-aerobic configurations in the activated sludge process; anaerobic treatment only; chemical, mechanical, and thermal methods to cause cell disruption; the addition of chemicals that uncouple cellular energy utilization; the use of protozoa for bacteria consumption; and extreme SRT processes. A summary evaluation of the biomass reduction processes is given in Table 11-1. Extensive bench-scale testing was done to assess biomass reduction, and possible lower sludge production for the anaerobic-aerobic activated sludge process with enhanced biological phosphorus removal (EBPR). The other processes were evaluated based on available literature.

The proposed means for biomass reduction in anaerobic-aerobic EBPR process was COD loss in the anaerobic contact zone, but the exact mechanism for such loss was not determined in experimental work reported by previous investigators. This process was selected for further research in this study because it could easily be added to existing facilities, and provided other benefits such as better sludge settling and phosphorus removal without chemicals. Parallel reactor studies with and without the anaerobic contact zone were generally lacking in previous studies, but were a key element in the bench-scale evaluations done in this work. The possibility of a COD loss, and thus biomass reduction, was found to be very unlikely based on the bench-scale test and further evaluation of substrate transformation mechanisms for anaerobic fermentation reactors. The solids production between reactors operated with or without anaerobic contacting was similar for tests with synthetic and municipal wastewaters.

Anaerobic treatment instead of aerobic treatment processes are known to produce less biomass because of their basic metabolic reactions, which convert almost 90% of the degradable COD in the wastewater to methane. As shown in Table 11-1, a 60–80% reduction in biomass production is possible with anaerobic processes. Two issues associated with anaerobic treatment of low strength wastewaters at low to moderate temperatures are the potential to produce effluent with BOD and suspended solids concentrations that exceed secondary treatment standards, and the inability to maintain a sufficiently long SRT so that most of the influent degradable COD is converted to methane. Recent research results, however, indicate that these difficulties may be overcome through developments in reactor design. Bench-scale studies at 8–20 °C with different anaerobic reactor designs—including the anaerobic attached film expanded bed reactor

(AAFEBR), the anaerobic baffled reactor (ABR), the anaerobic filter (AF), and upflow anaerobic sludge blanket reactor (UASB)—showed that a secondary effluent quality could be achieved for municipal wastewater treatment. However, these processes are considered to be high-risk treatment alternatives until they can be demonstrated long-term in larger scale systems closer to full-scale design with variable operating conditions. A major concern is the long-term ability to control solids in these processes.

Results in the literature for bench-scale tests have shown that using an anaerobic reactor with membrane technology for the final liquid-solids separation is a feasible technology. However, this process needs to be demonstrated at pilot-plant or larger scale with sufficient results on the membrane performance to evaluate its long-term treatment potential, operating issues, and economic feasibility.

A number of sequential anaerobic-aerobic processes have been demonstrated to achieve a secondary treatment effluent quality and 40–60% less sludge production at bench-scale and small pilot-scale systems under warmer climates (temperatures ≥ 20 °C). In these processes 75–80% BOD removal was achieved in an anaerobic process, followed by fixed-film or activated sludge processes for effluent polishing. One literature source noted that the combined process was economically competitive with conventional aerobic treatment with sludge digestion for the warm climate application. The use of membrane separation may be competitive in cost to fixed-film or activated sludge polishing steps, but more pilot-scale test results are needed before a reasonable design can be developed for an anaerobic-membrane separation system to develop an economic analysis.

A number of cell disruption technologies were evaluated that cause cell lysis and can be incorporated into the RAS line of existing systems. The most extensive analysis is provided for disruption by ozonation, mechanical disintegration, and thermal treatment. Biomass reduction with the ozonation and thermal treatment processes was well documented, with possible biomass reductions in the range of 50–90% for ozonation and 20–80% for thermal treatment. Of these two processes, thermal treatment of RAS was least attractive due to the high energy cost to raise the mixed liquor to 80–100 °C, the greater space requirements, and the design and operational complexity. The ozonation process was shown to be feasible for both pilot and full-scale operations, with the amount of sludge reduction directly related to the ozone dose. However, a detailed cost analysis done in this study and also shown in literature cost analyses indicated that the ozonation process was at best only attractive for cases where solids processing and available disposal methods were extremely expensive, such as extensive hauling distances, or possibly incineration.

Cell disruption by mechanical disintegration is one of the least proven cell reduction technologies in this study. A process using a high-speed mixer (The KADY Biolysis process) showed potential biomass reduction in the range of 10–20%, but a controlled study proving actual reductions was lacking. However, at these reduction levels the reduction in sludge handling costs associated with the biomass reduction likely would not offset the energy and equipment cost for the disintegration process. More work at pilot and full-scale is needed to fully determine the potential for mechanical disintegration processes, but the energy demand for the process may be problematic.

Previous work has shown conclusively that the use of certain toxic chemicals, such as trichlorophenol and many others, can uncouple energy capture in bacteria cells and result in 50–60% biomass reduction. The cost of the chemical versus the savings in sludge handling costs has not been fully evaluated, as the major concern for the process is the discharge of the added toxic chemical in the plant effluent. It appears difficult to find a chemical that can cause cellular uncoupling that is not toxic. Thus, this technology has no potential application until a means is found to eliminate the toxic compound in the discharge.

The predator enhancement process has been shown in pilot-scale studies to be a feasible process for treatment of higher strength, highly soluble wastewaters. The main potential application for this process is thus in industrial wastewater treatment. It does require a more complex treatment system and more operational skill, but further evaluation of the process appears warranted in view of the degree of biomass reduction possible (50%).

The extreme SRT process, first evaluated over 30 years ago, is feasible today with the practical application of membrane bioreactors that can be used for high solids capture efficiency to maintain SRTs well above 50 days. The aerobic activated sludge treatment reactor would have to be quite large compared to more conventional approaches, which would result in unfavorable economics compared to other liquid treatment approaches. The long SRT might be better maintained in a sludge digestion process. The process may have some potential for onsite treatment systems where simple processes with minimal net sludge production are desired.

The following overall conclusions can be made from this study:

1. COD loss may not exist in anaerobic-aerobic EBPR systems and should not be included in the process design to predict less sludge production and less oxygen required.
2. Anaerobic treatment is the most promising biomass reduction technology and a variety of designs are possible to provide 40–80% less sludge production. Extensive pilot-scale testing of anaerobic-membrane separation processes and other anaerobic processes are needed to develop and evaluate feasible full-scale processes.
3. A predator enhancement process is available for 50% biomass reduction for moderate to high strength soluble industrial wastewaters. Site-specific pilot testing would be required prior to implementation.
4. Using ozone to cause cell disruption is very effective for reducing biomass (50–90% reduction), but the costs may limit its application only to situations where sludge disposal options have extreme costs, such as incineration.
5. Mechanical disintegration processes for cell disruption are not fully evaluated, but there should be concerns about the cost of operating the disintegration process versus the possible reduced sludge handling costs.
6. Chemical uncouplers, thermal treatment, and extreme SRT for biomass reduction during liquid treatment do not appear to be feasible processes due to concerns for effluent quality and/or costs.

Most biomass reduction technologies make biological phosphorus removal, and possibly biological nitrogen removal, more difficult compared to conventional biological treatment processes. For extensive biomass reduction levels, significant biological phosphorus removal will be impossible.

Table 11-1. Summary of Biomass Reduction Technologies Evaluation.

Biomass reduction process	Percent biomass reduction expected	Proven (Yes or No)	Systems used to demonstrate biomass reduction	Ease of retrofit for existing WWTPs	Operational impacts of process	Effect on treatment performance	Process risk
Anaerobic-Aerobic EBPR	20-25%	No	• Bench • Small pilot	Good	• Positive • provides P removal • improves sludge settling	• Positive • Nutrient removal process	Low
Anaerobic treatment only	60-80%	Yes	• Bench	Fair, site specific	limited experience for system operation	Additional steps needed for nutrient removal	High
Sequential anaerobic-aerobic	40-60%	Yes	• Bench • Pilot plant	Low-fair, site specific	limited experience for system operation	Additional steps needed for nutrient removal	•Moderate •No low temp. results
Cell disruption by ozonation	50-90%	Yes	• Bench • Pilot plant • Full scale	Good	Ozone generator and contactor operation	• Biological P removal not feasible • Nitrogen removal possible	• Low • High costs
Cell disruption by mechanical methods	10-20%	Yes	• Bench • Pilot plant • Full scale	Good	Increased complexity and maintenance	None	• Moderate • High costs
Cell disruption by thermal methods	20-80%	Yes	• Bench	Good	• Operated intensive • Very complex	• Biological P removal not feasible • Nitrogen removal possible	• Low • High costs
Chemical uncoupling	50-60%	Yes	• Bench	Good	Simple change	Produces toxic effluent	Moderate
Predator enhancement	10-80%	Yes	• Bench • Pilot plant	Moderate	• Increased operator skill • Increased monitoring • More complex	• Interferes with nutrient removal • Not successful with municipal	High
Extreme SRT	70-90%	Yes	• Bench	Low	• More complex	• Biological P removal not feasible • Nitrogen removal possible	Low

CHAPTER 12.0

RESEARCH NEEDS

Review of a wide range of technologies for biomass reduction revealed a number of research needs. These are listed below for each of the technologies and also for general issues that apply across all the technologies.

12.1 General Needs

- Development of a comprehensive cost analysis model would be useful for evaluating the tradeoff of the capital and operating costs for biomass reduction technologies versus the cost savings in sludge processing and disposal. Such a model would be useful for the evaluation of site specific applications for which retrofit approaches, sludge disposal/reuse, and unit costs vary.

- The fate of wastewater solids that are nonbiodegradable (nbVSS) in aerobic processes should be evaluated. What is the composition of these solids and are they amenable to biological hydrolysis under anaerobic conditions? Can they be consumed by higher life forms, such as protozoa and rotifers?

12.2 Anaerobic Processes

Anaerobic processes continue to offer great potential for wastewater treatment of low to moderate strength wastewaters. Key issues are long-term testing and demonstration of processes that appear to provide a secondary effluent quality and within that the ability to maintain influent solids and biomass at long SRTs. A number of process choices have been shown possible for anaerobic treatment including upflow anaerobic sludge blanket reactors, upflow expanded bed reactors, anaerobic baffled reactors, and anaerobic reactors with membrane separation, and anaerobic reactors with a subsequent aerobic polishing step. A goal of current proposed research may not be to determine which of these is best, but to determine how processes assure long SRTs for methanogenic populations, the fate of influent solids in these processes, and the ability of these processes to handle variable loadings. The potential for anaerobic treatment spans a wide range when municipal and industrial wastewaters are considered and temperatures are considered. Nevertheless, the technology evaluation suggests that anaerobic processes can be applied at lower wastewater strengths and temperatures than is current accepted practice. The Water Environment Research Foundation (WERF) has already sponsored a project to investigate membrane separation processes in anaerobic systems treating low to moderate strength wastewater. Some key research needs that apply to the anaerobic processes are listed below.

- Long-term demonstration of anaerobic processes at sufficient scale under realistic real world operating conditions is needed.
- Previous studies do not indicate a need for alkalinity addition, but this should be a concern for pH control in view of the higher CO_2 content in the gas phase of anaerobic processes. A protocol to evaluate the effect of influent alkalinity and wastewater strength, and the anaerobic system operating conditions is needed to better understand pH control and the need for alkalinity addition. Alkalinity addition may represent a significant operatin cost in anaerobic processes.
- Process schemes should be developed and tested to provide nutrient removal in anaerobic systems used to treat municipal wastewaters. Some processes have indicated the potential for integration of nitrification and denitrification within the anaerobic treatment process. Other ideas may be developed to provide some combination of biological and chemical processes for phosphorus removal.
- A better fundamental understanding of anaerobic degradation kinetics for both hydrolysis fermentation and methanogenesis is needed for low temperature operations.

12.3 Anaerobic-Aerobic EBPR Processes

The ability of anaerobic-aerobic EBPR processes to produce a COD loss is questionable based on the results of research done in this study. Further research in this area will require more sophisticated analytical tools to follow carbon substrates and the development and rigorous evaluation of possible COD loss mechanisms. For example, work with radio-labeled carbon compounds might provide an experimental procedure to produce more conclusive results.

12.4 Cell Disruption by Mechanical Methods

The use of mechanical methods of cell disruption may be limited by the energy needed versus the degree of biomass reduction and related cost savings. However, it may be possible to apply this technology at a lower energy level to develop dispersed biological solids that can be used in a predator enhancement process.

- Determine the ability to couple a mechanical disintegration process at lower energy with a predator enhancement process.

12.5 Sonification Processes

Sonification has not been applied in biological liquid treatment processes, but the technology has shown good results for solids disruption and enhanced sludge digestion rates.

- Evaluate the use of sonification to minimize biomass reduction in activated sludge processes.

12.6 Ozonation Process

The ozonation process was shown to be a feasible and effective biomass reduction technology, but its cost might limit its application to more unique situations. However, the previous research did not address the possibility of applying a greater wastewater flow to an existing system to take advantage of the lower net solids yield and thus the increased capacity of the aeration tank. Of course, other overall process considerations, such as aeration capacity and classification capacity, would have to be considered. The process may also aid biological nutrient removal.

- Determine the effect of the ozonation process on the required SRT for BOD removal and nitrification in comparison to processes without RAS ozonation.
- Determine if the ozonation process can also enhance biological phosphorus removal and denitrification in biological nutrient removal systems, in addition to biomass reduction.

12.7 Chemical Uncouplers

Chemical uncouplers are a proven means of biomass reduction but the cost of the uncoupling chemicals and effluent toxicity is of concern. The process may have more potential and be more manageable if applied to aerobic digestion processes. A smaller volume would result for the treatment of the toxic chemical.

- Investigate the use of chemical uncouplers in the aerobic digestion processes to accelerate solids destruction rates and to produce a lower final residual mass.
- Investigate methods to biodegrade the toxic compound used to cause chemical uncoupling.

12.8 Predator Enhancement Process

The predator enhancement process has been shown to be a feasible process for moderate to high strength soluble industrial wastewaters. Further research may increase the range of application of this process.

- Determine the fate of nonbiodegradable volatile suspended solids (nbVSS) in predator enhancement processes. Can the higher life forms consume nonbiological colloidal solids?
- Determine if physical or chemical methods can be used to economically create dispersed solids sufficient for predator growth in a predator enhancement process.

12.9 Extreme SRT Processes

While extreme SRT processes require relatively larger aeration detention times and membrane separation, they may be attractive for small on-site wastewater treatment applications where minimal sludge production is desired. Previous research did not fully determine the mechanism(s) for biomass reduction and the role of higher life forms, such as protozoa and rotifers.

- Determine the role of higher life forms and operating conditions that promote minimal biomass and solids production in extreme SRT process.

APPENDIX A

RANKING OF BIOMASS REDUCTION TECHNOLOGIES

A decision analysis was used to rank the biomass reduction technologies. The criteria for the evaluation and ranking of the technologies are shown in Table A-1.

Table A-1. Criteria and Weighting Values for Evaluation and Ranking of Biomass Reduction Technologies.

Criteria	Weight
I. COST EFFECTIVENESS	
Sludge Trmt & Disp. Cost	10
Plant Capacity Impacts	4
II. CERTAINTY/ RELIABILITY	
Level of Demonstration	3
Quality of Evidence	6
Process Stability	5
III. IMPLEMENTATION IMPACTS	
Biomass Reduction	10
Ease of Implementation	3
Impact on O/M	7
Impact on Plant Performance	8

For each biomass reduction technology a score was applied at each of the criteria listed in Table A-1. The score for each of the criteria could range from 0-10, and tables were developed for each category to define the basis for the 0-10 rating. These are defined as utility factors and their values are shown for each category Table A-2 to A-4. The total score for a given technologies was then equal to the sum of the utility factor times the weighting factor of each criteria.

Table A-2. Cost Effectiveness Utility Values.

Sludge Treatment & Disposal Cost, $/DT WAS Reduction	Utility Value
800-1000	0
600-800	2
400-600	4
200-400	7
100-200	9
0-100	10
Plant Capacity Impacts	
10	0
8	2
6	5
4	8
2	10

Table A-3. Certainty/Reliability Utility Values.

Level of Demonstration	Utility Value
None	0
Bench	3
Pilot	5
Full-scale	8
Full-scale Long-term	10
Quality of Evidence	**Utility Value**
Conceptual	0
Anecdotal (no ctrl)	1
Test, Quantified, No Ctrl (w/o FFP)	3
Test, Quantified, No Ctrl (w/FFP)	5
Control, side by side, QA/QC	8
The CACILLAC	10
Process Stability	**Utility Value**
Process Failure Possible	0
Process Upsets Temporary	3
Process Upsets Possible but Avoidable	6
No Known Upset Factors	8
Bullet-proof	10

Table A-4. Implementation Factors Utility Values.

Biomass Reduction Capability, %	Utility Value	Ease of Implementation	Utility Value
0-10	2	Reconstruct Plant	0
10-20	5	Internal w/Mods to Existing	4
20-50	8	Internal w/Additions	8
50-80	9	Minimal (External Changes)	10
80-100	10		
Impact on Operations and Maintenance	**Utility Value**	**Impact on Plant Performance**	**Utility Value**
High skill, high time	2	Certain NOV	0
Intense oversight	3	Potential NOV	2
Mod. time & greater skill	5	Neg. impacts BOD/SS	7
Existing staff, more time	8	Neg. impacts Nutrient Rem.	8
Existing staff-minimal impact	10	No Impact	10

The cost effectiveness was based on a qualitative assessment and represents the range of impacts of cost for implementing the biomass reduction technology and cost savings.

The final rankings were done under three modes in view of the uncertainty in the costs and in the actual biomass reduction efficiency, which could affect plant capacity. In Table A-5 the scores and rankings for the three conditions are summarized and represent the accumulative scores of the workshop participants.

Table A-5. Summary of Workshop Scores for the Biomass Reduction Technology Evaluation and Ranking.

With all costs		Without capacity impacts
Anaerobic-Aerobic, AO BioP	423	Bioconversion
Anaerobic Processes	394	Anaerobic-Aerobic, AO BioP
Bioconversion	384	Long SRT
Long SRT	378	Anaerobic Processes
Ozonation	362	Ozonation
Cell Disruption	323	Cell Disruption
Uncoupling	318	Uncoupling
Predator Enhancement	300	Predator Enhancement
Anaerobic-Aerobic, Sludge Ferm.	252	Anaerobic-Aerobic, Sludge Ferm.
Without costs		
Ozonation	322	
Long SRT	308	
Anaerobic Processes	304	
Bioconversion	294	
Anaerobic-Aerobic, AO BioP	293	
Cell Disruption	273	
Predator Enhancement	230	
Uncoupling	228	
Anaerobic-Aerobic, Sludge Ferm.	162	

APPENDIX B

Ozonation Process Cost Analysis

Biomass Reduction by Ozonation of Return Activated Sludge

A cost analysis was performed by CH2M-HILL engineers to assess the cost-effectiveness of retrofitting an existing plant with the ozonation process described in Chapter 6.0 for biomass reduction. The cost of implementing an ozone biomass reduction process into an existing activated sludge plant was compared to the cost savings in solids processing and disposal/reuse. The costs for the ozone system included capital and operating cost processing, while the cost savings for solids were based strictly on operating costs reductions.

The cost analyses were done for a municipal and industrial wastewater scenario at an influent flowrate of 5.0 Mgal/day (18,925 m^3/d). Two solids handling scenarios were evaluated, termed Option A and Option B. Option A consisted of a conventional solids processing scheme of centrifuge thickening anaerobic digestion, centrifuge dewatering, and transport and reuse by land application. Option B assumed that the waste sludge could be simply dewatered by centrifugation and disposed at a sanitary landfill. The municipal wastewater treatment process consisted of primary and secondary activated sludge treatment with the production of both primary and secondary waste sludge. The industrial wastewater scenario was for a moderate strength industrial wastewater (BOD = 600 mg/L), which contained mainly soluble organic material with little nonbiodegradable volatile suspended solids. Thus no primary treatment was used and the solids produced consisted of mostly waste activated sludge biomass. Thus the biomass reduction process would have a greater affect on the total solids reduction for the industrial wastewater treatment versus the municipal wastewater treatment. The municipal and industrial wastewater treatment volumes are illustrated in Figures B-1 and B-2 with the Option A and Option B solids processing methods.

The wastewater characteristic for the municipal wastewater treatment application is given in Table B-1, and the design summaries for the primary and secondary treatment processes are given in Table B-2 and B-3 respectively. A typical influent BOD and TSS concentration of 200 mg/L was assumed. For primary treatment typical BOD and TSS removal efficiencies of 35 and 65%, respectively were assumed. The solids production from primary treatment is this 5,000 lb (2270 kg/d) TSS/day.

The secondary activated sludge process was designed with a 10-day SRT, 3,000 mg/L MLSS concentration, and a net solids yield of 0.72 g TSS/g BOD removed. The total secondary sludge produced is 3,500 lb/day (1590 kg/d) with about 60% of that or 1960 lb/day (890 kg/d) as biomass. For the ozone retrofit application, (the key design parameters for the ozone process retrofit are shown in Table B-4. The biomass reduction was arbitrarily selected at 60%, which fist well within the range of biomass reduction demonstrated in previous test work (Chapter 6.0). For the example design this is equal to 1,180 lb/day (536 kg/d) biomass destruction by the ozone process. Based on previously reported results, an ozone dose of 0.2 g O_3/g VSS destroyed was assumed, requiring a 240 lb/day (109 kg/d) on-site ozone generator. A 15-minute contact time was assumed for the ozonation of the RAS sidestream.

The influent wastewater conditions for the industrial WWTP are summarized in Table B-5. The BOD concentration was set at 600 mg/L with most of it as soluble organics.

The industrial WWTP secondary process design assumptions and design values for the ozonation process are summarized in Table B-6 and B-7, respectively. An SRT of 10 days was used, and the net solids yield was 0.44 g TSS/g BOD_r with about 90% of the solids being biomass. For this higher strength wastewater the total solids production is 11,000 lb TSS/day (5000 kg/d) and 10,000 lb/d (4550 kg/d) of biomass. For this application 60% biomass reduction was also assumed for the ozonation retrofit.

For the cost analysis two detailed cost estimate tasks were completed. The first was to determine the cost in \$/dry ton of solids for processing the waste sludge via Option A or Option B. This analysis did not require estimates of capital cost, as this cost analysis scenario was for a retrofit application and thus the processing equipment and facilities were already present. The sludge processing cost determined the costs for power, chemicals, labor and maintenance for each processing step. The costing was done using the total amount of waste sludge without the ozonation process, and then using the total operating costs and sludge quantity to arrive at a total \$/dry ton cost for sludge processing. This was done separately for both the municipal and industrial WWTP scenarios.

The sludge processing costs are shown in Table B-8. Key costing parameters are shown in Table B-9. They include the interest rate (6%) for amortization, a power cost at 6¢/kwh, natural gas cost at 9¢/therm, and estimating percentages of capital costs for site work, yard piping, contingency, and general operations and maintenance.

The operating costs for in \$/dry ton are given at the top of Table B-8 for each step of sludge processing; pumping, thickening, digestion, and dewatering for Option A and only pumping and dewatering for Option B. Land application fees for reuse and landfill fees for disposal in a sanitary landfill for Options A and B, respectively are added to processing costs. These fees are very site-specific and the cost estimate was done using representative values from the Denver, CO WWTP, which was considered to be at moderate levels. The analysis in Table B-8 also determined sludge hauling cost in \$/dry ton-mile.

Tables B-10 and B-11 summarize the costing of the ozonation process installation and operation for the municipal and industrial WWTP applications, respectively. The equipment and installation costs for the ozone system were based on the required ozonation capacity, which was higher for the industrial WWTP, which had a much greater amount of biomass reduction. The operating costs included RAS pumping, additional aeration energy for oxidation of the cell material released from ozonation, power for the ozone generator, and purchase of pure oxygen for feed to the ozone generator. Maintenance and labor costs were included for the ozone equipment. The capital cost was amortized for a 10-year period at 6% interest to obtain a daily cost that could be added to the daily operating costs. The cost of ozonation versus the reduced cost for sludge processing for Options A and B for the municipal WWTP application are compared in Tables B-12 and B-13, respectively. For Option A at 60% biomass reduction, the quantity of sludge that is destroyed by ozonation is 1200 lb/day (545 kg/d). The sludge processing cost is $315/ton per Table B-7, so the daily savings in processing costs is $189. However, the total ozone cost for amortization and operation is $782/day or a difference of $543/day. If the sludge hauling cost for the solids destroyed is greater than $543/day, the hauling cost was not included in the sludge processing costs in Table B–11 so that the ozonation process could be an economically viable alternative. At the estimated hauling cost of $1.4/ton-mile, the distance would have to exceed 646 miles. For a similar analysis for Option B the distance would have to exceed 828 miles. The bottom row in Tables B-12 and B-13 also show that the sludge processing cost would have to be $1230/ton to make the ozonation process competitive if the hauling distance is 50 miles. A comparison of the ozonation sludge processing cost savings for biomass reduction is similarly compared in Tables B-14 and B-15 for the industrial WWTP application for Options A and B, respectively. The hauling distance needed to make the ozonation process feasible is again excessive, at 271 miles for Option A and 436 miles for Option B, respectively. For a 50-mile hauling distance the sludge processing cost would have to be greater than $2030/ton to make the ozonation process attractive. The fact that the processing cost for the industrial WWTP application has to be higher in $/ton than for the municipal application even though A had a greater fraction of the solids destroyed by ozonation, indicates that the conventional sludge processing methods are less costly than the cost of processing the sludge by ozonation to reduce the net amount of sludge processed.

I. Conventional Municipal Wastewater Treatment Design

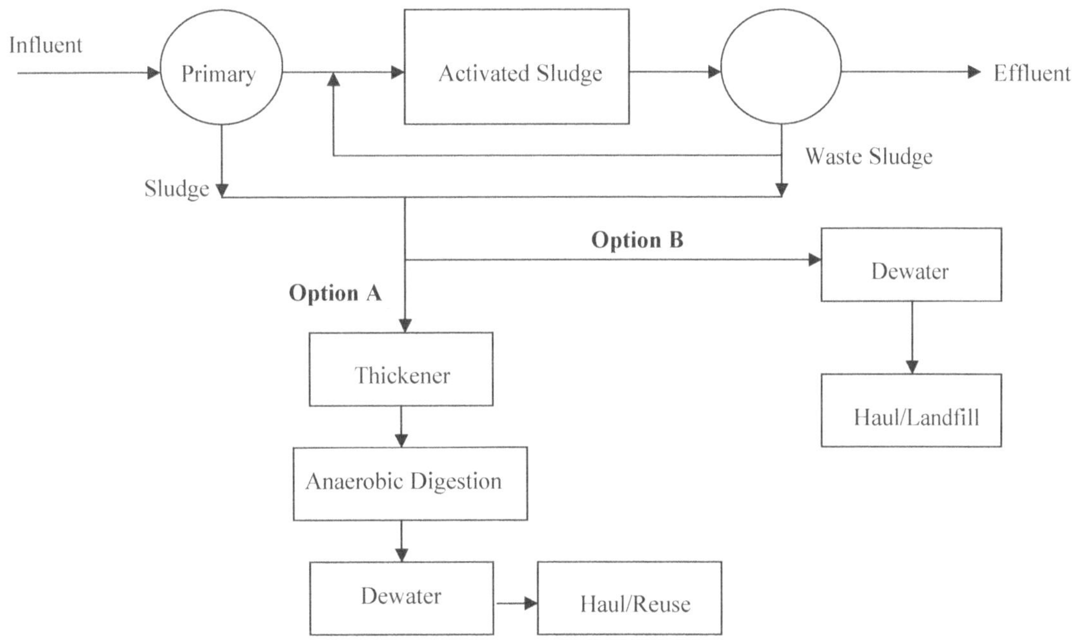

Figure B-1. Treatment Scheme and Sludge Processing Alternatives for Municipal Wastewater Treatment with Primary Clarification.

II. Industrial Wastewater Treatment

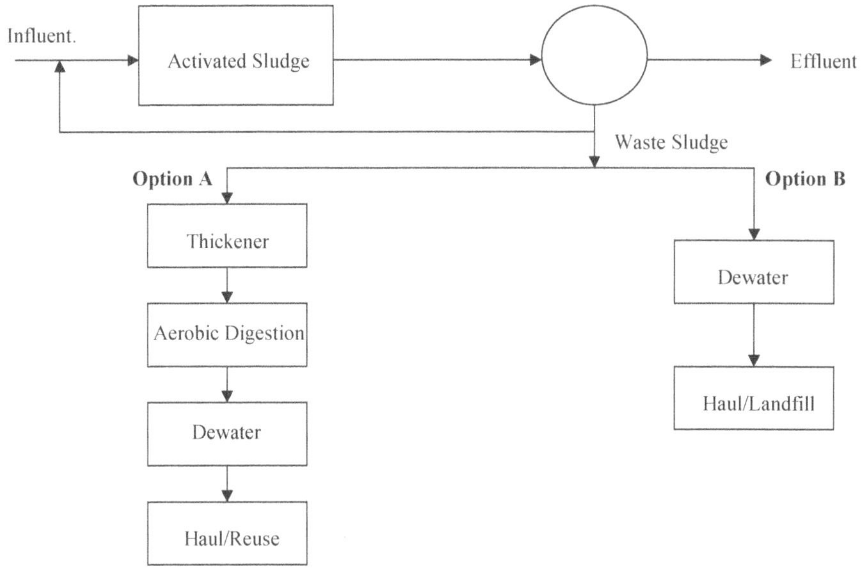

Figure B-2. Treatment Scheme and Sludge Processing Alternatives for Industrial Wastewater Treatment of Highly Soluble Wastewater.

Table B-1. Influent Wastewater Design Conditions for Cost Analysis for Municipal Wastewater Treatment.

Parameter	**Units**	**Value**
Flow	Mgal/day	5.0
BOD	mg/L	200
TSS	mg/L	200
TKN	mg/L	35
Temperature	^{0}C	15

Table B-2. Summary of Municipal Wastewater Treatment Plant Primary Treatment Process Design for 5 Mgal/day (18,925 m³/d) Flow and Influent Wastewater Characteristics Shown in Table A-1.

Design Parameter	**Units**	**Value**
Primary Treatment		
overflow rate	gpd/ft2	1000
2 primary clarifiers	diameter, ft	56.0
TSS Removal	percent	65.0
BOD Removal	percent	35.0
solids production	lb/day	5000
Primary Effluent		
BOD	mg/L	130
TSS	mg/L	80
TKN	mg/L	30

Table B-3. Summary of Municipal Wastewater Treatment System Secondary Activated Sludge Process Design for 5 Mgal/day (18,925 m^3/d) Flow and Influent Wastewater Characteristics Shown in Table A-2.

Design Parameter	Units	Value
Process assumptions		
SRT	days	10.0
MLSS	mg/L	3,000
nbVSS yield coefficient	g nbVSS/g BOD applied	0.32
synthesis yield coefficient	g TSS/g BODr	0.70
endogenous decay coefficient	g TSS/g VSS-d	0.08
net solids yield	g TSS/g BODr	0.72
net biomass yield	g TSS/g BODr	0.40
nitrogen used in cell synthesis	g N/g TSS	0.12
fine bubble actual aeration efficiency	percent	12.0
secondary clarifier overflow rate	gpd/ft2	600
Design summary for secondary		
Aeration tank volume	Mgal	1.4
Aeration detention time	hrs	6.8
Total O2 req'd including nitrification	lb/hr	442
aeration rate	scfm	3,570
blower energy	Hp	180
2 secondary clarifiers diameter	ft	73.0
Total solids wasted	lb/day	3,500
biomass wasted	lb/day	1,960
Solids production summary		
Total primary plus secondary	lb/day	8,500
secondary	lb/day	3,500
biomass	lb/day	1,960

Table B-4. Design Assumptions for Biomass Reduction with Ozonation Process for Municipal Wastewater Treatment Example.

Item	Assumptions	Tank Additions	Equipment Added	Aeration Oxygen Needed	Biomass Reduction, lb/day
Ozone Contact tank	10% of RAS @15 min contact time	700 ft^3.			
Ozone generator	0.20 g O_3 per g VSSr		240 lb/day O_3 generator		
Pumping Recycle sludge			1-2.5 Hp		
Process oxygen change				1,600 lb/d	
Biomass Reduction	60% reduction				1,180

Table B-5. Influent Wastewater Design Conditions for Cost Analysis for Industrial Wastewater Treatment.

Parameter	Units	Value
Flow	Mgal/day	5.0
BOD	mg/L	600
TSS	mg/L	100
TKN	mg/L	40
Temperature	^{0}C	15

*sufficient phosphorus assumed in wastewater

Table B-6. Summary of Industrial Wastewater Treatment System Secondary Activated Sludge Process Design for 5 Mgal/day Flow and Influent Wastewater Characteristics Shown in Table A-5.

Design Parameter	Units	Value
Process assumptions		
SRT	days	10.0
MLSS	mg/L	2,500
nbVSS yield coefficient	g nbVSS/g BOD applied	0.04
synthesis yield coefficient	g TSS/g BODr	0.70
endogenous decay coefficient	g TSS/g VSS-d	0.08
net solids yield	g TSS/g BODr	0.44
net biomass yield	g TSS/g BODr	0.40
nitrogen used in cell synthesis	g N/g TSS	0.12
fine bubble actual aeration efficiency	percent	12.0
secondary clarifier overflow rate	gpd/ft2	600
Design summary for secondary		
Aeration tank volume	Mgal	4.44
Aeration detention time	hrs	21.3
Total O2 req'd (assumes no nitrification)	lb/hr	1,180
aeration rate	scfm	9,530
blower energy	Hp	480
2 secondary clarifiers diameter	ft	73.0
Total solids wasted	lb/day	11,000
biomass wasted	lb/day	10,000
Solids production summary		
Total secondary	lb/day	11,000
biomass	lb/day	10,000

Table B-7. Design Assumptions for Biomass Reduction with Ozonation Process for Industrial Wastewater Treatment Example.

Item	Assumptions	Tank Additions	Equipment Added	Aeration Oxygen Needed	Biomass Reduction, lb/day
Ozone Contact tank	10% of RAS @15 min contact time	700 ft^3.			
Ozone generator	0.20 g O_3 per g VSSr		1,200 lb/day O_3 generator		
Pumping Recycle sludge			1-2.5 Hp		
Process oxygen change				8,400 lb/day	
Biomass Reduction	60% reduction				6,000

Table B-8. Sludge Processing and Disposal/Reuse Cost Estimates for Municipal and Industrial WWTP Scenarios.

1.1 WERF 00-CTS-10 **O&M Cost Backup—Solids**	
Parameter	**Value**
Solids Processing and Disposal Cost—Summary	
costs below as $$/dry ton of raw sludge.	
Municipal WWTP	**Cost/Dry Ton**
Sludge pumping:	
To thickening centrifuge	$3.70
To anaerobic digester	$1.70
To dewatering centrifuge	$2.50
Total sludge pumping	$7.90
Centrifuge thickening	$100.40
Anaerobic digestion	$42.20
Centrifuge dewatering—Option A	$93.80
Centrifuge dewatering—Option B	$129.80
Subtotal operating cost (w/o disposal)—Option A	$244
Subtotal operating cost (w/o disposal)—Option B	$133
Land application (reuse)	$71
MWRD, based on $118/DT of digested sludge, converted to raw sludge equiv	
Landfill tip fee	$10
Assuming 30% solids, $33/wet ton, based on recent experience	
Total unit solids cost for Option A—without hauling	$315
Total unit solids cost for Option—without hauling	$143
Industrial WWTP	
Sludge pumping:	
To thickening centrifuge	$5.50
To anaerobic digester	$2.50
To dewatering centrifuge	$2.20
Total sludge pumping	$10.20
Centrifuge thickening	$162.10
Anaerobic digestion	$37.70
Centrifuge dewatering—Option A	$85.00
Centrifuge dewatering—Option B	$119.70
Subtotal operating cost (w/o disposal)—Option A	$295
Subtotal operating cost (w/o disposal)—Option B	$125

Land application (reuse)	$71
MWRD, based on $118/DT of digested sludge, converted to raw sludge equivalent	
Landfill tip fee	$10
Assuming 30% solids, $33/wet ton, based on recent experience	
Total unit solids cost for Option A—without transport	$366
Total unit solids cost for Option B—without transport	$135
Solids Processing and Disposal Unit Costs	
Assumptions:	
Municipal sec sludge production, lb/day	3,500
Equivalent reference plant size	4
Based on 200 mg/L BOD, 30% primary removal, 0.75 yield	
Municipal primary sludge production, lb/day	5,000
Assumes 200 mg/L TSS, 60% removal based on above plant size	
Total municipal sludge production, lb/day	8,500
Industrial secondary sludge production, lb/day	11,000
Industrial primary sludge production, lb/day	0
Total industrial sludge production, lb/day	11,000
Sludge Pumping (to centrifuge thickener)	
Municipal Wastewater Treatment	
Sludge quantity, lb/day	3,500.00
Unthickened sludge quantity, gal/day @ 1% solids from secondary	41,966
Pumping capacity, gpm	29.1
Energy required, hp (based on 30 ft TDH, 60% Eff.	0.368
Total power requirement, kw	0.3
Total annual power usage, kwh	2,405
Unit power cost, $/kwh	$0.06
Annual power cost	$144
Annual labor hours	85
Per "Capital and O&M Cost Estimates for Attached Growth	
Biological Wastewater Treatment Processes", EPA (1989), Figure C-12	
Annual labor cost—Based on $23/hr plus 30% burden	$2,553
Annual maintenance materials cost	$3,029
"Capital and O&M Cost Estimates for Attached Growth	
Biological Wastewater Treatment Processes", EPA (1989),. Figure C-14	
Total annual labor, power and materials cost	$5,726

Industrial Wastewater Treatment	
Sludge quantity, lb/day	11,000.00
Unthickened sludge quantity, gal/day—1% solids for secondary sludge	131,894
Pumping capacity, gpm	91.6
Energy required, hp—Based on 30 ft TDH, 60% efficiency	1.156
Total power requirement, kw	0.9
Total annual power usage, kwh	7,558
Unit power cost, $/kwh	$0.06
Annual power cost	$453
Annual labor hours	137
Annual labor cost—Based on $23/hr plus 30% burden	$4,101
Annual maintenance materials cost	$6,532
Total annual labor, power and materials cost	$11,086
Sludge Pumping	
Municipal Wastewater Treatment	
Sludge quantity, lb/day—Secondary sludge only	3,500.00
Unthickened sludge quantity, gal/day—Based on net 4%solids	10,492
Pumping capacity, gpm	7.3
Energy required, hp -Based on 30 ft TDH, 60% efficiency	0.092
Total power requirement, kw	0.1
Total annual power usage, kwh	601
Unit power cost, $/kwh	$0.06
Annual power cost	$36
Annual labor hours	48
Annual labor cost—Based on $23/hr plus 30% burden	$1,438
Annual maintenance materials cost	$1,194
Total annual labor, power and materials cost	$2,669
Industrial Wastewater Treatment	
Sludge quantity, lb/day—Secondary sludge only	11,000.00
Unthickened sludge quantity, gal/day—Based on net 4% solids	32,974
Pumping capacity, gpm	22.9
Energy required, hp—Based on 30 ft TDH, 60% efficiency	0.289
Total power requirement, kw	0.2
Total annual power usage, kwh	1,889
Unit power cost, $/kwh	$0.06
Annual power cost	$113

Annual labor hours	77
Annual labor cost—Based on $23/hr plus 30% burden	$2,310
Annual maintenance materials cost	$2,576
Total annual labor, power and materials cost	$5,000
Sludge Pumping A40	
Municipal Wastewater Treatment	
Sludge quantity, lb/day—Based on 50% VS reduction, 80% VS	5,100.00
Unthickened sludge quantity, gal/day—Based on net 3% solid	20,384
Pumping capacity, gpm—Based on 30 ft TDH, 60% efficiency	14.2
Energy required, hp	0.179
Total power requirement, kw	0.1
Total annual power usage, kwh	1,168
Unit power cost, $/kwh	$0.06
Annual power cost	$70
Annual labor hours	63
Annual labor cost—Based on $23/hr plus 30% burden	$1,893
Annual maintenance materials cost	$1,865
Total annual labor, power and materials cost	$3,829
Industrial Wastewater Treatment	
Sludge quantity, lb/day -Based on 50% VS reduction, 80% VS	6,600.00
Unthickened sludge quantity, gal/day—Based on net 3% solids	26,379
Pumping capacity, gpm	18.3
Energy required, hp—Based on 30 ft TDH, 60% efficiency+A136	0.231
Total power requirement, kw	0.2
Total annual power usage, kwh	1,512
Unit power cost, $/kwh	$0.06
Annual power cost	$91
Annual labor hours	70
Annual labor cost—Based on $23/hr plus 30% burden	$2,106
Annual maintenance materials cost	$2,218
Total annual labor, power and materials cost	$4,415
Centrifuge Thickening (WAS)	
Municipal Wastewater Treatment	
Sludge quantity, lb/day—Secondary sludge only	3,500.00
Unthickened sludge quantity, gal/day—Based on net 1% solids	41,966
Centrifuge feed, gpm	29.1

Power requirement, hp	10.2
EPA Process Design Manual, "Sludge Treatment and Disposal" P. 5-59, 0.35 hp/gpm cap	
Total power requirement, kw	7.6
Total annual power usage, kwh	66,658
Unit power cost, $/kwh	$0.06
Annual power cost	$3,999
Annual labor hours	3,496
EPA Process Design Manual, "Sludge Treatment and Disposal" Figure 5-25	
Annual labor cost—Based on $23/hr plus 30% burden	$104,865
Annual polymer cost—Based on 3 lb/dry ton @ $2.50/lb	$4,791
Annual maintenance material cost	$42,109
EPA Process Design Manual, "Sludge Treatment and Disposal" Figure 5-26	
Total annual labor, power and materials cost	$155,765
Industrial Wastewater Treatment	
Sludge quantity, lb/day—Secondary sludge only	11,000.00
Unthickened sludge quantity, gal/day—Based on net 1% solids	131,894
Centrifuge feed, gpm	91.6
Power requirement, hp	32.1
EPA Process Design Manual, "Sludge Treatment and Disposal" P. 5-59, 0.35 hp/gpm cap	
Total power requirement, kw	23.9
Total annual power usage, kwh	209,496
Unit power cost, $/kwh	$0.06
Annual power cost	$12,570
Annual labor hours	7,742
EPA Process Design Manual, "Sludge Treatment and Disposal" Figure 5-25	
Annual labor cost—Based on $23/hr plus 30% burden	$232,263
Annual polymer cost—Based on 3 lb/dry ton @ $2.50/lb	$15,056
Annual maintenance material cost	$65,463
A34Per EPA Process Design Manual, "Sludge Treatment and Disposal" Figure 5-26	
Total annual labor, power and materials cost	$325,352
Anaerobic Digestion	
Municipal Wastewater Treatment	
Sludge quantity, lb/day Includes primary and secondary	8,500.00
Thickened sludge quantity, gal/day—Based on net 4% solids	25,480
Sludge heating power, BTU/hr	313,438
EPA Process Design Manual, "Sludge Treatment and Disposal" Figure 6-32	
Makeup heating requirement, BTU/hr	125,375

EPA C145 Manual, "Sludge Treatment and Disposal", Figure 6-32	
Annual gas requirement, Therms—Assumes 70% boiler efficiency	54,914.30
Annual gas cost—Assumes no reuse of digester gas	$7,194
Mixing power requirement, BTU/hr	42,500.00
EPA Process Design Manual, "Sludge Treatment and Disposal" Figure 6-32, 240,000 BTU/dry ton	
Mixing power requirement, hp	16.7
Total power requirement, kw	12.4
Total annual power usage, kwh	109,057
Unit power cost, $/kwh	$0.06
Annual power cost	$6,543
Equivalent plant size, mgd—Based on 1880 lb/day/mgd	4.5
Total annual labor and materials cost	$51,729
EPA Process Design Manual, "Sludge Treatment and Disposal"	
Figure 6-34 (using 1900 lb/day/mgd to obtain equiv plant mgd)	
Total annual labor, power and materials cost	$65,466
Industrial Wastewater Treatment	
Sludge quantity, lb/day—Includes primary and secondary	11,000.00
Thickened sludge quantity, gal/day—Based on net 4% solids	32,974
Sludge heating power, BTU/hr	405,625
EPA Process Design Manual, "Sludge Treatment and Disposal" Figure 6-32	
Makeup heating requirement, BTU/hr	162,250
EPA Process Design Manual, "Sludge Treatment and Disposal" Figure 6-32	
Annual gas requirement, Therms—Assumes 70% boiler efficiency	71,065.50
Annual gas cost—Assumes no reuse of digester gas	$9,310
Mixing power requirement, BTU/hr	55,000.00
EPA Process Design Manual, "Sludge Treatment and Disposal"	
Figure 6-32, 240,000 BTU/dry ton	
Mixing power requirement, hp	21.6
Total power requirement, kw	16.1
Total annual power usage, kwh	141,133
Unit power cost, $/kwh	$0.06
Annual power cost	$8,468
Equivalent plant size, mgd—Based on 1880 lb/day/mgd	5.9
Total annual labor and materials cost	$57,938
EPA Process Design Manual, "Sludge Treatment and Disposal"	
Figure 6-34 (using 1900 lb/day/mgd to obtain equiv plant mgd)	
Total annual labor, power and materials cost	$75,715

Centrifuge Dewatering—Option A	
Municipal Wastewater Treatment	
Sludge quantity, lb/day—Based on 50% VS reduction, 80% VS	5,100.00
Digested sludge quantity, gal/day—Based on net 2.5% solids	24,460
Centrifuge feed, gpm	17
Power requirement, hp	5.9
EPA C184 Manual, "Sludge Treatment and Disposal" P. 5-59, 0.35 hp/gpm cap	
Total power requirement, kw	4.4
Total annual power usage, kwh	38,852
Unit power cost, $/kwh	$0.06
Annual power cost	$2,331
Annual labor hours	2,579
EPA Process Design Manual, "Sludge Treatment and Disposal, Figure 5-25	
Annual labor cost—Based on $23/hr plus 30% burden	$77,371
Annual polymer cost	$31,646
Based on information from MWRD (Denver, CO)—$34/dry ton	
Annual maintenance material cost	$34,202
EPA Process Design Manual, "Sludge Treatment and Disposal" Figure 5-26	
Total annual labor, power and materials cost	$145,549
$/ton??	39.9
Industrial Wastewater Treatment	
Sludge quantity, lb/day—Based on 50% VS reduction, 80% VS	6,600.00
Digested sludge quantity, gal/day—Based on net 2.5% solids	31,655
Centrifuge feed, gpm	22
Power requirement, hp	7.7
EPA Process Design Manual, "Sludge Treatment and Disposal" P. 5-59, 0.35 hp/gpm cap	
Total power requirement, kw	5.7
Total annual power usage, kwh	50,279
Unit power cost, $/kwh	$0.06
Annual power cost	$3,017
Annual labor hours	2,965
EPA Process Design Manual, "Sludge Treatment and Disposal" Figure 5-25	
Annual labor cost—Based on $23/hr plus 30% burden	$88,960
Annual polymer cost	$40,953
Based on information from MWRD (Denver, CO)—$34/dry ton	
Annual maintenance material cost	$37,774
EPA Process Design Manual, "Sludge Treatment and Disposal" Figure 5-26	
Total annual labor, power and materials cost	$170,703

Centrifuge Dewatering—Option B	
Municipal Wastewater Treatment	
Sludge quantity, lb/day Based on raw sludge	8,500.00
Unthickened sludge quantity, gal/day -Based on net 2.% solids	40,767
Centrifuge feed, gpm	28.3
Power requirement, hp	9.9
EPA Process Design Manual, "Sludge Treatment and Disposal" P. 5-59, 0.35 hp/gpm cap	
Total power requirement, kw	7.4
Total annual power usage, kwh	64,753
Unit power cost, $/kwh	$0.06
Annual power cost	$3,885
Annual labor hours	3,435
EPA Process Design Manual, "Sludge Treatment and Disposal" Figure 5-25	
Annual labor cost—Based on $23/hr plus 30% burden	$103,048
Annual polymer cost	$52,743
Based on information from MWRD (Denver, CO—$34/dry ton	
Annual maintenance material cost	$41,642
EPA Process Design Manual, "Sludge Treatment and Disposal" Figure 5-26	
Total annual labor, power and materials cost	$201,317
Industrial Wastewater Treatment	
Sludge quantity, lb/day Based on raw sludge	11,000.00
Unthickened sludge quantity, gal/day—Based on net 2.5 percent solids	52,758
Centrifuge feed, gpm	36.6
Power requirement, hp	12.8
EPA Process Design Manual, "Sludge Treatment and Disposal" P. 5-59, 0.35 hp/gpm cap	
Total power requirement, kw	9.6
Total annual power usage, kwh	83,798
Unit power cost, $/kwh	$0.06
Annual power cost	$5,028
Annual labor hours	4,031
EPA Process Design Manual, "Sludge Treatment and Disposal" Figure 5-25	
Annual labor cost—Based on $23/hr plus 30% burden	$120,937
Annual polymer cost	$68,255
Based on information from MWRD (Denver, CO)—$34/dry ton	
Annual maintenance material cost	$45,991
EPA Process Design Manual, "Sludge Treatment and Disposal" Figure 5-26	
Total annual labor, power and materials cost	$240,211

Sludge Hauling	
Municipal Wastewater Treatment	
Sludge quantity, lb/day	5100
sludge, tons/day	2.6
Sludge quantity, cu yd/day At 20% solids	14.4
Truckloads/day Assumes 15 cu yd truck	1
Transport unit cost:	
Truck cost per mile	
Maintenance	$0.37
from "Estimating Water Treatment Costs", EPA, 1979, Table 168	
Fuel Based on 3 mpg, $1.5/gal	$0.50
Total	$0.87
Labor cost per mile	
Hour per mile—Assumes average 30 mph	0.0333
Labor cost per mile—Based on $23/hr plus 30% burden	$1.00
Total haul cost per mile of distance to site—15 miles	$3.74
Total cost per year for haul distance = 5 miles	$6,559
Total cost per year for haul distance = 10 miles	$13,118
Total cost per year for haul distance = 20 miles	$26,236
Total cost per year for haul distance = 30 miles	$39,355
30 miles $/ton for hauling	$42.28
Industrial Wastewater Treatment	
Sludge quantity, lb/day	6600
Sludge quantity, cu yd/day—At 20% solids	18.7
Truckloads/day -Assumes 15 cu yd truck	1.2
Transport unit cost:	
Truck cost per mile	
Maintenance	$0.37
from "Estimating Water Treatment Costs", EPA, 1979, Table 168	
Fuel—Based on 3 mpg, $1.5/gal	$0.50
Total	$0.87
Labor cost per mile	
Hour per mile—Assumes average 30 mph	0.0333
Labor cost per mile—Based on $23/hr plus 30% burden	$1.00
Total haul cost per mile of distance to site	$3.74
Total cost per year for haul distance = 5 miles	$8,488
Total cost per year for haul distance = 10 miles	$16,977

Total cost per year for haul distance = 20 miles	$33,953
Total cost per year for haul distance = 30 miles	$50,930
cost in $/ton/mile	1.4

Table B-9. General Assumptions Used in Cost Analyses.

1.1.1 WERF 00-CTS-10		
General Parameters		
Parameter	**Value**	**Comments**
Cost parameters		
Discount rate	6.00%	
Power cost, $/kwh	$0.06	
Natural gas cost, $/therm	$0.09	
Sitework allowance	10.0%	
Yard piping allowance	10.0%	
Contingency	30.0%	
General O&M cost	3.0%	Does not include consumables (e.g., chemicals, membranes) or power cost
Replacement cost	2.0%	Annual cost for sinking fund to replace equipment and facilities as they wear out.
Engineering & Administrative	15.0%	
CCI values:		
Current	6397	October 2001, per ENR

Table B-10. Estimated Cost of Ozonation Process Retrofit to Municipal WWTP Including Capital and Operating Costs.

Facility/Component	Size/Capacity	Cost
Capital Costs		
Ozone contact tank,cu ft	700	$17,366
"Estimating Water Treatment Costs", EPA, 1979, Figure 10		
Ozone generation package, lb/hr	10	$115,446
"Estimating Water Treatment Costs", EPA, 1979, Figure 9		
Building, sq ft	1500	$225,000
Assume size per previous project; cost at $150/sq ft		
Recycle pump	1 @ 2.5 hp	$6,000
Assume $6,000 each installed per previous projects		
Subtotal Construction Cost		$363,812
Electrical and I&C—15% of subtotal construction cost		$54,572
Sitework allowance—10% of subtotal construction cost		$36,381
Yard piping allowance—15% of subtotal construction cost		$54,572
Total Construction Cost		**$509,336**
Contingency		$152,801
Grand Total Construction Cost		**$662,137**
Engineering, Legal, Administration		**$99,321**
Total Capital Cost		**$761,458**
Amortized Annual Capital Cost (10 yrs)		**$103,458**
Amortized Annual Capital Cost (20 yrs)		**$66,387**
Operating Costs		
Annual costs:		
Equipment power:		
Recycle pump	1@2.5 hp	2.5
Additional oxygen needed, lb/day	1600	26.7
Total horsepower		29.2
Ozone system kw		85.1

"Estimating Water Treatment Costs", EPA, 1979, Figure 11		
Building energy		1.2
"Estimating Water Treatment Costs", EPA, 1979, Figure 11		
Total power requirement, kw		108.1
Total annual power usage, kwh		947,000
Unit power cost, $/kwh		$0.06
Annual power cost		$56,820
Ozone system, capacity in lb/day	240	
Oxygen, lb/day@8lb O3:100 lb O2, cost at $100/ton	3000	$54,750
Maintenance material		$10,156
"Estimating Water Treatment Costs", EPA, 1979, Figure 11		
Labor		$22,201
"Estimating Water Treatment Costs", EPA, 1979, Figure 12		
General O&M cost		$19,864
Subtotal Annual Cost		**$163,792**
Solids handling and disposal cost (reduction)—Option A		($63,261)
Total Annual Cost —Option A		**$100,531**
Sludge reduction, lb/day	1100	
Solids handling and disposal cost (reduction)—Option B		($28,781)
Total Annual Cost —Option B		**$135,010**
Assumed Discount Rate, %		6%
Present Worth of Annual Costs (10 years)		**$1,205,521**
Present Worth of Annual Costs (20 years)		**$1,878,678**
Sludge haul cost reduction per year		
Total cost per year for haul distance = 5 miles		$1,415
Total cost per year for haul distance = 10 miles		$2,829
Total cost per year for haul distance = 20 miles		$5,659
Total cost per year for haul distance = 30 miles		$8,488

Table B-11. Estimated Cost of Ozonation Process Retrofit to Industrial WWTP Including Capital and Operating Costs.

Facility/Component	Size/Capacity	Cost
Capital Costs		
Ozone contact tank,cu ft	700	$17,366
"Estimating Water Treatment Costs", EPA, 1979, Figure 10		
Ozone generation package, lb/hr	50	$324,950
"Estimating Water Treatment Costs", EPA, 1979, Figure 9		
Building, sq ft	2000	$300,000
Recycle pump	1 @ 2.5 hp	$6,000
Subtotal Construction Cost		$648,316
Electrical and I&C—15% of subtotal construction cost		$97,247
Sitework allowance—10% of subtotal construction cost		$64,832
Yard piping allowance—15% of subtotal construction cost		$97,247
Total Construction Cost		**$907,642**
Contingency		$272,293
Grand Total Construction Cost		**$1,179,935**
Engineering, Legal, Administration		**$176,990**
Total Capital Cost		**$1,356,925**
Amortized Annual Capital Cost (10 yrs)		**$184,363**
Amortized Annual Capital Cost (20 yrs)		**$118,303**
Operating Costs		
Annual costs:		
Equipment power:		
Recycle pump	1@2.5 hp	2.5
Additional oxygen needed, lb/day	8400	140
Total horsepower		142.5
Ozone system kw		364.7
"Estimating Water Treatment Costs", EPA, 1979, Figure 11		
Building energy		4
"Estimating Water Treatment Costs", EPA, 1979, Figure 11		
Total power requirement, kw		475.1

Total annual power usage, kwh		4,161,655
Unit power cost, $/kwh		$0.06
Annual power cost		$249,699
Ozone system, capacity in lb/day:	1200	
Oxygen, lb/day@8lb O3:100 lb O2, cost at $100/ton	15000	$273,750
Maintenance material		$27,940
"Estimating Water Treatment Costs", EPA, 1979, Figure 11		
Labor		$46,554
"Estimating Water Treatment Costs", EPA, 1979, Figure 12; 1500 hr/yr		
General O&M cost		$35,398
Subtotal Annual Cost		**$633,341**
Solids handling and disposal cost (reduction)—Option A		($345,058)
Total Annual Cost —Option A		**$288,284**
Sludge reduction, lb/day	6000	
Solids handling and disposal cost (reduction)—Option B		($156,989)
Total Annual Cost —Option B		**$476,353**
Assumed Discount Rate, %		6%
Present Worth of Annual Costs (10 years)		**$4,661,449**
Present Worth of Annual Costs (20 years)		**$7,264,377**
Sludge haul cost reduction per year		
Total cost per year for haul distance = 5 miles		$7,717
Total cost per year for haul distance = 10 miles		$15,433
Total cost per year for haul distance = 20 miles		$30,866
Total cost per year for haul distance = 30 miles		$46,300

Table B-12. Summary of Cost Analysis for Municipal Wastewater with Ozonation Biomass Reduction Process. Biomass Reduction Equal 60% or 1180 lb/day for 5 Mgal/day WWTP. Option A Sludge Processing—Thickening, Digestion, Dewatering, Land Application.

Item	Ozonation Costs	Savings for Sludge Processing	Cost Difference
Sludge reduction		0.60 ton/day.	
sludge processing costs Option A		$315/ton	
Savings in processing		$189/day	
Ozone capital amortization costs	$283/day		
Ozone operation costs	$449/day		
Total Ozone costs	$782/day		
Ozonation costs minus sludge processing cost savings			$543/day
Biosolids hauling costs			$1.4/Ton/mile
Hauling distance for breakeven			646 miles
Processing cost in $/Ton to breakeven at 50 mi. haul distance		$1230/Ton	

Table B-13. Summary of Cost Analysis for Municipal Wastewater with Ozonation Biomass Reduction Process. Biomass Reduction Equal 60% or 1180 lb/day for 5 Mgal/day WWTP. Option B Sludge Processing—Thickening, Dewatering, Landfill.

Item	Ozonation Costs	Savings for Sludge Processing	Cost Difference
Sludge reduction		0.60 ton/day.	
sludge processing costs Option A		$143/ton	
Savings in processing		$86/day	
Ozone capital amortization costs	$283/day		
Ozone operation costs	$449/day		
Total Ozone costs	$782/day		
Ozonation costs minus sludge processing cost savings			$696/day
Biosolids hauling costs			$1.4/Ton/mile
Hauling distance for breakeven			828 miles
Processing cost in $/Ton to breakeven at 50 mi. haul distance		$1230/Ton	

Table B-14. Summary of Cost Analysis for Industrial Wastewater with Ozonation Biomass Reduction Process. Biomass Reduction Equal 60% or 6000 lb/day for 5 Mgal/day WWTP. Option A Sludge Processing—Thickening, Digestion, Dewatering, Land Application.

Item	Ozonation Costs	Savings for Sludge Processing	Cost Difference
Sludge reduction		3 ton/day.	
sludge processing costs Option A		$366/ton	
Savings in processing		$1098/day	
Ozone capital amortization costs	$505/day		
Ozone operation costs	$1735/day		
Total Ozone costs	$2240/day		
Ozonation costs minus sludge processing cost savings			$1142/day
Biosolids hauling costs			$1.4/Ton/mile
Hauling distance for breakeven			271 miles
Processing cost in $/Ton to breakeven at 50 mi. haul distance		$2030/Ton	

Table B-15. Summary of Cost Analysis for Industrial Wastewater with Ozonation Biomass Reduction Process. Biomass Reduction Equal 60% or 6000 lb/day for 5 Mgal/day WWTP. Option B Sludge Processing—Thickening, Dewatering, Landfill.

Item	Ozonation Costs	Savings for Sludge Processing	Cost Difference
Sludge reduction		3 ton/day.	
sludge processing costs Option A		$135/ton	
Savings in processing		$405/day	
Ozone capital amortization costs	$505/day		
Ozone operation costs	$1735/day		
Total Ozone costs	$2240/day		
Ozonation costs minus sludge processing cost savings			$1835/day
Biosolids hauling costs			$1.4/Ton/mile
Hauling distance for breakeven			436 miles
Processing cost in $/Ton to breakeven at 50 mi. haul distance		$2030/Ton	

APPENDIX C

ANAEROBIC MEMBRANE SEPARATION PILOT PLANT TEST PROGRAM

Background

The review of biomass reduction technologies showed that one of the most effective ways to reduce biomass production is to use an anaerobic process for wastewater treatment instead of an aerobic process. However, the literature review on anaerobic processes in Chapter 2.0 showed that there were very few reports of anaerobic treatment of low-strength industrial wastewaters and municipal wastewaters at temperatures of 20°C or below. A review of previous studies on anaerobic treatment processes (Chapter 2.0) suggested in only few studies that anaerobic treatment of low strength wastewaters at low temperature using fixed film and sludge blanket treatment processes may be feasible. No results for treating low strength wastewaters at low temperature were found for Membrane Coupled Anaerobic Bioreactor processes (MCAB). The MCAB treatment process offers advantages over the other processes that should lead to more efficient, stable, and reliable anaerobic treatment of municipal wastewaters at lower temperatures. The availability of membrane separation provides an opportunity to achieve high solids capture from the treated effluent, and thus the potential to maintain a long solids retention time (SRT) and relatively high methanogenic population in the anaerobic reactor. The ability to maintain a longer SRT should help overcome biokinetic limitations for anaerobic treatment for low or medium strength wastewaters at low temperatures. An additional advantage is that the MCAB is capable of achieving six law orders of magnitude coliform bacteria removal, Key issues for the successful application of a MCAB process are:

- Can membrane fouling be controlled to provide adequate operating flux, acceptable cleaning methods, and acceptable operating conditions following the anaerobic reactor?
- Can an anaerobic process be employed with solids recycle from the membrane separation step to maintain a longer SRT for methanogens and thus a high degree of organic conversion to methane in the anaerobic step?

The main objectives for the pilot plant testing are to determine if a MCAB process, treating municipal wastewater, can be operated at temperatures of 20°C to consistently achieve a secondary effluent quality with a majority of the COD conversion occurring in the anaerobic reactor zone. Once the process performance is established at 200°C, a lower temperature operation will be tried, provided that there is time available within the planned study period ending in May, 2004. The temperature will be decreased at 2°C increments over two-week time periods. If the gas production and treatment efficiency is maintained, the temperature reduction will continue towards the ambient temperatures at the site, which is expected to be 15-16°C.

Anaerobic Pilot Plant Description

A MCAB process proposed for this study is shown in Figure C-1 below and consists of a four-staged upflow anaerobic baffled reactor, followed by a short aerobic zone and membrane liquid-solids separation. Solids captured from the membrane separator are first sent to an anaerobic completely mixed reactor before recycle to the first or second upflow anaerobic reactor.

The staged anaerobic upflow baffled reactor is expected to provide a number of advantages, based on results of previous studies:

- Staged reactors provide more efficient biological degradation kinetics.
- The upflow anaerobic reactors have the potential to develop a sludge blanket with a very high concentration of active anaerobic solids, in the range of 2-10% solids, which can provide a long SRT and minimize treatment reactor volume requirements.
- The upflow anaerobic reactor is expected to provide good solids capture to minimize the downstream solids concentration and loads.

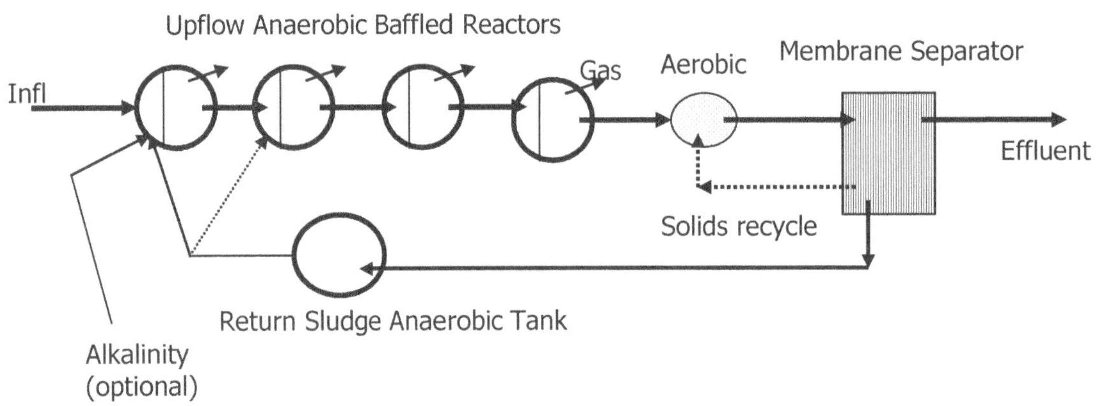

Figure C-1. Schematic of Anaerobic Baffled Reactor Membrane Separation Pilot Plant Process.

The short-detention time aerobic step in the proposed treatment scheme was added to provide conditions in which good membrane liquid-solids separation has been demonstrated previously. Prior work with only anaerobic effluents fed to membrane separation has shown more potential for membrane fouling due to dispersed bacteria. The aerobic step will also provide further treatment of the anaerobic effluent, which could have a higher soluble BOD concentration than that for conventional aerobic processes, used for low strength wastewaters, due to the presence of volatile fatty acids.

However, as one of the goals of the proposed process is to maximize the methanogenic organism population, the effect of aerobic exposure on the anaerobic population is of interest. Previous work in our labs has shown that methanogens can withstand oxygen exposure and resume degradation activities, but we don't know if the time of exposure in the aerobic process will affect the viability of the methanogens. In view of this concern, an anaerobic holding tank will be used to receive the returned solids from the aerobic step and membrane separation. This tank will provide time for the development and recovery of active methanogenic organisms before returning this recycle stream to the anaerobic baffled reactor. The continual seeding of methanogens to the ABR should assist the process efficiency.

A pilot plant has been installed at the West Point wastewater treatment plant in Seattle, WA and plant personnel will operate it. The pilot plant tank volumes are approximately 370 gallons per anaerobic reactor (total of 1480 gallons), a 150-gallon aerobic tank, and a 250-gallon anaerobic solids return tank. The membrane separation unit is provided by U.S. Filter and is rated for flows from 2-5 gallons/min. A 3 mm screen is installed between the ABR effluent and the aeration tank prior to the membrane separation unit. The pilot plant will be operated initially at 2 gpm for nominal detention times of 3 hours/anaerobic stage (12 hours total) and an aerobic time of 1.25 hours. The aerobic stage SRT will be maintained at 2-3 days. The test flows are expected to be increased to 4 gpm later in the study, which will reduce these detention times by one half. The pilot plant operation is expected to occur over a six-month period.

The ABR will be started with West Point anaerobic digesters seed diluted to a concentration of about 1% and added to 1/3 of the volume of the first three ABR stages. The influent will be heated to a temperature of 20 ^{o}C during start up and alkalinity will be added to maintain the pH in the digesters near 7.0. The aerobic reactor will be operated with an acceptable SRT for the membrane separation operation, but will be started with low SRTs (1-4 days) to minimize the aerobic holding time of methanogens.

Figure C-2. Pilot Plant ABR Tanks in Series Inside Trailer.

Pilot Plant Testing Objectives

The basic aim of the pilot study is to test the feasibility of the ABR/aerobic/membrane separation (AAM) process in the treatment of municipal wastewater at temperatures at and below 20 °C. COD and solids balances will be performed for the process to assess its treatment characteristics and performance. The key questions to be addressed by the testing are provided below:

Treatment Performance Issues

- What degree of COD conversion occurs overall in the ABR?
- What is the effluent TSS/VSS concentration and soluble BOD from the ABR fed to the aerobic unit?
- What is the effluent quality of the membrane permeate?

Solids Production Issues

- What is the net solids production for the MCAB process in gTSS/g CODr?
- What is the fraction of influent COD converted to methane gas in the ABR?
- What is the methanogenic activity of the anaerobic holding reactor return sludge?

Process Operation and Characteristic Issues

- What are the solids blanket characteristics of the ABR and what is the solids distribution between ABR stages?
- How does the COD removal activity vary in each stage of the ABR?
- What is the percent methane and CO_2 in the ABR gas production?
- What flux and pressure drop can be maintained with the membrane separation unit with the recommend operation and cleaning methods?

To meet these objectives the following sampling and analyses program is planned with King County personnel. Effluent quality, gas production and methane content, BOD and COD removal, and solids balances will be possible with this data collection program.

The final report on The Anaerobic Membrane Separation Pilot Plant Test Program will be available on the Water Environment Research Foundation's website (http://www.werf.org/) upon completion.

Sample Location	ABR Influent	ABR #1	ABR #2	ABR #3	ABR #4	ABR EFFl.	Aerobic Reactor	Membrane Reactor (not really a reactor)	Membrane Return	Membrane Permeate	Anaerobic Dig Tank
Analysis											
BOD	D									P	
TCOD	D					D					
sCOD	D					D				P	
TSS/VSS	D	1/wk	1/wk	1/wk		D	D		D		1/wk
VFA		3/wk	3/wk	3/wk		3/wk					3/wk
Alkalinity	D					D					
pH		D	D	D	D		D				D
Gas production		C	C	C	C						D
%CO2		3/wk	3/wk	3/wk	3/wk						3/wk
%CH4		P	P	P	P						P
Solids blanket depth		P	P	P	P						
mixed sludge conc.		P	P	P	P						
Turbidity										D	
Pressure drop								D		D	
ORP		P	P	P	P						

c- continuous

P- at least once per week/TBD

REFERENCES

Ahn, K.H., K.Y. Park, S.K. Maeng, J.H. Hwang, J.W. Lee, K.G. Song, and S. Choi (2002a). Ozonation of wastewater sludge for reduction and recycling. *Water Science and Technology*, Vol. 46, No. 10, pp. 71-77

Ahn, KH., I.T. Yeom, K.Y. Park, S.K. Maeng, Y. Lee, K.G. Song, and J.H. Hwang (2002b). Reduction of sludge by ozone treatment and production of carbon source for denitrification. *Water Science and Technology*, Vol. 46, No. 11-12, pp. 121-125.

Albertson, O.E. (1987). The control of bulking sludge: from the early innovators to current practice. *Journal Water Pollution Control Federation*, Vol. 59, No. 4, pp.172-179.

Alderman, B.J., T.L. Theis and A.G. Collins (1998). Optimal design for anaerobic pretreatment of municipal wastewater. *Journal of Environmental Engineering-ASCE* **,** Vol. 124, No. 1, pp. 4-10.

Angenent, L.T. and R.R. Dague (1996). *Initial studies on the anaerobic migrating blanket reactor*. 51st Purdue Industrial Waste Conference, Ann Arbor Press, Inc., Chelsea, MI, pp 271-288.

Angenent, L.T. and S.W. Sung (2001). Development of anaerobic migrating blanket reactor (AMBR), a novel anaerobic treatment system. *Water Research*, Vol. 35, No. 7, pp. 1739-1747.

Angenent, L.T., G.C. Banik and S.W. Sung (2001). Anaerobic migrating blanket reactor treatment of low-strength wastewater at low temperatures. *Water Environment Research*, Vol. 73, No. 5, pp. 567-574.

Angenent, L.T., D.D. Zheng, S.H. Sung, and L. Raskin (2002). Microbial community structure and activity in a compartmentalized, anaerobic bioreactor. *Water Environment Research*, Vol. 74, No. 5, pp. 450-???.

APHA (1998). *Standard Methods for the Examination of Water and Wastewater*, 20th ed. American Public Health Association, Washington, D.C.

Bachmann, A., V.L. Beard, and P.L. McCarty (1983). Comparison of fixed film reactors with a modified sludge blanket reactor. Fixed Film Biological Processes for Wastewater Treatment. Y.C. Wu and E.D. Smith. NJ, Noyes Data.

Bachmann, A., V.L. Beard, and P.L. McCarty (1985). Performance-Characteristics of the Anaerobic Baffled Reactor. *Water Research*, Vol. 19, No. 1, pp.99-106.

Bagley, D.M. and T.S. Brodkorb (1999). Modeling microbial kinetics in an anaerobic sequencing batch reactor–Model development and experimental validation. *Water Environment Research*, Vol. 71, No. 7, pp. 1320-1332.

Ballapragada, B. (1996). *Reductive Dechlorination of chlorinated Ethenes by Mixed Anaerobic Consortia*. Ph.D. Thesis, University of Washington. Seattle, WA.

Banik, G.C. and R.R. Dague (1996). ASBR treatment of dilute wastewater at low temperatures. *69th Annual Water Environmental Conference*, USA, pp 235-246.

Barber, W.P. and D.C. Stuckey (1998). The influence of start-up strategies on the performance of an anaerobic baffled reactor. *Environmental Technology*, Vol. 19, No. 5, pp. 489-501.

Barber, W.P. and D.C. Stuckey (1999). The use of the anaerobic baffled reactor (ABR) for wastewater treatment: A review. *Water Research*, Vol. 33, No. 7, pp. 1559-1578.

Barker, P.S. and P.L. Dold (1995). COD and nitrogen mass balances in activated sludge systems. *Water Research*, Vol. 29, No. 2, pp. 633-643.

Barker, P.S. and P.L. Dold (1997). General model for biological nutrient removal in activated sludge systems: model presentation. *Water Environment Research*, Vol. 69, No. 3, pp. 1-18.

Barnard, J.L. (1974). Cut P and N without chemicals. *Water and Wastes Engineering*, Vol. 11, pp. 41-44.

Bizier, P.A. (1999). Evaluation of a biosolids minimization system. In *Proceedings of the ASCE CSCE National Conference on Environmental Engineering*, July 25-28, 1999, Norfolk, VA. pp. 397-406.

Bodik, I., B. Herdova, and M. Drtil (2002a). The use of upflow anaerobic filter and AnSBR for wastewater treatment at ambient temperature. *Water Research*, Vol. 36, No. 4, pp. 1084-1088.

Bodik, I., K. Kratochvil, B. Herdova, G. Tapia and E. Gasparikova (2002b). Municipal wastewater treatment in the anaerobic-aerobic baffled filter reactor at ambient temperature. *Water Science and Technology*, Vol. 46, No. 8, pp.127-135.

Boopathy, R. and D.M. Sievers (1991). Performance of a Modified Anaerobic Baffled Reactor to Treat Swine Waste. *Transactions of the ASAE*, Vol. 34, No. 6, pp. 2573-2578.

Boopathy, R. and A. Tilche (1991). Anaerobic-Digestion of High-Strength Molasses Waste-Water Using Hybrid Anaerobic Baffled Reactor. *Water Research,* Vol. 25, No. 7, pp. 785-790.

Boopathy, R. and A. Tilche (1992). Pelletization of Biomass in a Hybrid Anaerobic Baffled Reactor (Habr) Treating Acidified Waste-Water. *Bioresource Technology*, Vol. 40, No. 2, pp.101-107.

Boopathy, R., V.F. Larsen, and E. Senior (1988). Performance of Anaerobic Baffled Reactor (Abr) in Treating Distillery Wastewater from a Scotch Whiskey Factory. *Biomass*, Vol. 16, No. 2, pp. 133-143.

Bordacs, K. and K. Tracy (1988). Retention of organic storage products in anaerobic-aerobic activated sludge. *Proceedings of the 61st Annual Conference of the Water Pollution Control Federation,* Dallas, Texas

Camacho, P., S. Deleris, V. Geaugey, P. Ginestet, and E. Paul (2002). A comparative study between mechanical, thermal and oxidative disintegration techniques of waste activated sludge. *Water Science and Technology*, Vol. 46, No. 10, pp. 79-87.

Camacho, P., V. Geaugey, P. Ginestet, and E. Paul (2002). Feasibility study of mechanically disintegrated sludge and recycle in the activated-sludge process. *Water Science Technology*, Vol. 46, No. 10, pp. 97-104.

Camp Dresser & McKee (2000). *Report: Evaluation of the KADY Biolysis System at the Springfield Regional Wastewater Treatment Facility*, July 2000.

Cech, J.S. and P. Hartman (1990) Glucose induced break down of enhanced biological phosphate removal. *Environmental Technology*, Vol. 11, pp.651-656.

Cech, J.S. and P. Hartman (1993) Competition between polyphosphate and polysaccharide accumulating bacteria in enhanced biological phosphate removal systems. *Water Research*, Vol. 27, pp. 1219-1225.

Chen, G.H., H.K. Mo, S. Saby, W.K. Yip, and Y. Liu (2000). Minimization of activated sludge production by chemically stimulated energy spilling. *Water Science and Technology*, Vol. 42, No. 12, pp. 189-200.

Chen, G.H., H.K. Mo, and Y. Liu (2002). Utilization of a metabolic uncoupler, 3,3', 4 ', 5-tetrachlorosalicylanilide (TCS) to reduce sludge growth in activated sludge culture. *Water Research*, Vol. 36, No. 8, pp. 2077-2083.

Chernicharo, C.A.L. and M.D. Cardoso (1999). Development and evaluation of a partitioned upflow anaerobic sludge blanket (UASB) reactor for the treatment of domestic sewage from small villages. *Water Science and Technology*, Vol. 40, No. 8, pp. 107-113.

Choo, K.H. and C.H. Lee (1996a). Effect of anaerobic digestion broth composition on membrane permeability. *Water Science and Technology*, Vol. 34, No. 9, pp. 173-179.

Choo, K.H. and C.H. Lee (1996b). Membrane fouling mechanisms in the membrane-coupled anaerobic bioreactor. *Water Research*, Vol. 30, No. 8, pp. 1771-1780.

Choo, K.H. and C.H. Lee (1998). Hydrodynamic behavior of anaerobic biosolids during crossflow filtration in the membrane anaerobic bioreactor. *Water Research*, Vol. 32, No. 11, pp. 3387-3397.

Choo, K.H. and H.D. Stensel (2000). Sequencing batch membrane reactor treatment: nitrogen removal and membrane fowling evaluation. *Water Environment Research*, Vol. 72, No. 4, pp. 490-498.

Christensson, M. and T. Welander (2001). Decreasing the biosludge production by applying the low sludge production (LSP) process on municipal wastewater—a pilot study. Lund Sweden, ANOX AB.

Collins, A.G., T.L. Theis, S. Kilambi, L. He, and S.G. Pavlostathis (1998). Anaerobic treatment of low-strength domestic wastewater using an anaerobic expanded bed reactor. *Journal of Environmental Engineering-ASCE,* Vol. 124, No. 7, pp. 652-659.

Cox, H.H.J. and M.A. Deshusses (1999). Biomass control in waste air biotrickling filters by protozoan predation. *Biotechnology and Bioengineering*, Vol. 62, No. 2, pp. 216-224.

Crites, R. and G. Tchobanoglous (1998). *Small and Decentralized Wastewater Management Systems*, McGraw-Hill.

Dague, R.R., G.C. Banik, and T.G. Ellis (1998). Anaerobic sequencing batch reactor treatment of dilute wastewater at psychrophilic temperatures. *Water Environment Research*, Vol. 70, No. 2, pp. 155-160.

Deleris, S., V. Geaugey, P. Camacho, H. Debellefontaine, and E. Paul (2002). Minimization of sludge production in biological processes: an alternative solution for the problem of sludge disposal. *Water Science and Technology*, Vol. 46, No. 10, pp. 63-70.

De Man, A.W.A., P.C. Grin, R. Roersma, K.C.F. Grolle, and G. Lettinga (1986). Anaerobic treatment of sewage at low temperatures. *Anaerobic Treatment, a Grown-up Technology,* Amsterdam, NL, pp. 451-466.

De Man, A.W.A., G.B.J. Rijs, G. Lettinga, and W. Starkenburg (1988a). Anaerobic treatment of sewage using a granular sludge bed UASB reactor. *Fifth International Symposium on Anaerobic Digestion*, Bologna, Italy, pp. 735-738.

De Man, A.W.A., A.R.M. van der Last, and G. Lettinga (1988b). The use of EGSB and UASB anaerobic systems for low-strength soluble and complex wastewaters at temperatures ranging from 8-30°C. *Proceedings of the International Symposium on Anaerobic Digestion*, Bologna, Italy, pp. 197.

Derycke, D. and W. Verstraete (1986). Anaerobic treatment of domestic wastewater by a pilot scale polyurethane carrier reactor. *Anaerobic Treatment, a Grown-up Technology*, Amsterdam, September 16-18, pp. 437-450.

Egemen, E., J. Corpening, J. Padilla, R. Brennan, and N. Nirmalakhandan (1999). Evaluation of ozonation and cryptic growth for biosolids management in wastewater treatment. *Water Science Technology*, Vol. 39, No. 10-11, pp.155-158.

El-Gohary, F.A., F.A. Nasr, and H.I. Aly (1999). Cost-effective pretreatment of food-processing industrial wastewater. *Water Science and Technology*, Vol. 40, No. 7, pp. 17-24.

Elmitwalli, T.A., J. Soellner, A. De Keizer, H. Bruning, G. Zeeman, and G. Lettinga (2001). Biodegradability and change of physical characteristics of particles during anaerobic digestion of domestic sewage. *Water Research*, Vol. 35, No. 5, pp. 1311-1317.

Elmitwalli, T., G. Zeeman, and G. Lettinga (2001). Anaerobic treatment of domestic sewage at low temperature. *Water Science and Technology*, Vol. 44, No. 4, pp. 33-40.

Elmitwalli, T.A., K.L.T. Oahn, G. Zeeman, and G. Lettinga (2002a). Treatment of domestic sewage in a two-step anaerobic filter/anaerobic hybrid system at low temperature. *Water Research*, Vol. 36, No. 9, pp. 2225-2232.

Elmitwalli, T.A., V. Sklyar, G. Zeeman, and G. Lettinga (2002b). Low temperature pre-treatment of domestic sewage in an anaerobic hybrid or an anaerobic filter reactor. *Bioresource Technology*, Vol. 82, No. 3, pp. 233-239.

Foster, D. (1992). Cell disruption: Breaking up is hard to do. *Biotechnology*, Vol. 10, pp. 1539-1541.

Fox, P. and V. Venkatasubbiah (1996). Coupled anaerobic/aerobic treatment of high-sulfate wastewater with sulfate reduction and biological sulfide oxidation. *Water Science and Technology*, Vol. 34, No. 5-6, pp. 359-366.

Frankin, R.J. (2001). Full-scale experiences with anaerobic treatment of industrial wastewater. *Water Science and Technology*, Vol. 44, No.8, pp. 1-6.

Fuchs, W., H. Binder, G. Mavrias, and R. Braun (2003). Anaerobic treatment of wastewater with high organic content using a stirred tank reactor coupled with a membrane filtration unit. *Water Research*, Vol. 37, No. 4, pp. 902-908.

Garuti, G., M. Dohanyos, and A. Tilche (1992). Anaerobic-Aerobic Combined Process for the Treatment of Sewage with Nutrient Removal–the Ananox®Process. *Water Science and Technology*, Vol. 25, No. 7, pp. 383-394.

Gaudy, A.F., P.Y. Yang, and A.W. Obayashi (1971). Studies on the total oxidation of activated sludge with and without hydrolytic treatment. *Journal of Water Pollution Control Federation*, Vol. 43, No. 1, pp. 40-52.

Ghyoot, W. and W. Verstraete (1999). Reduced sludge production in a two-stage membrane-assisted bioreactor. *Water Research*, Vol. 34, No. 1, pp. 205-215.

Goncalves, R.F., V.L. de Araujo, and V.S. Bof (1999). Combining upflow anaerobic sludge blanket (UASB) reactors and submerged aerated biofilters for secondary domestic wastewater treatment. *Water Science and Technology*, Vol. 40, No. 8, pp. 71-79.

Grant, S. and K.C. Lin (1995). Effects of Temperature and Organic Loading on the Performance of Upflow Anaerobic Sludge Blanket Reactors. *Canadian Journal of Civil Engineering*, Vol. 22, No. 1, pp. 143-149.

Grin, P.C., R. Roersma, and G. Lettinga (1983). Anaerobic treatment of raw sewage at lower temperatures. *European Symposium on Anaerobic Wastewater Treatment*, Noordwijkerhout, NL, pp. 335-648.

Grobicki, A. and D.C. Stuckey (1989). The Role of Formate in the Anaerobic Baffled Reactor. *Water Research*, Vol. 23, No. 12, pp. 1599-1602.

Grobicki, A. and D.C. Stuckey (1991). Performance of the Anaerobic Baffled Reactor under Steady-State and Shock Loading Conditions. *Biotechnology and Bioengineering*, Vol. 37, No. 4, pp. 344-355.

Guiot, S.R. and L. Vandenberg (1985). Performance of an Upflow Anaerobic Reactor Combining a Sludge Blanket and a Filter Treating Sugar Waste. *Biotechnology and Bioengineering*, Vol. 27, No. 6, pp. 800-806.

Hernandez, A.E., L.C. Belalcazar, M.S. Rodriguez, and E. Giraldo (2002). Retention of granular sludge at high hydraulic loading rates in an anaerobic membrane bioreactor with immersed filtration. *Water Science and Technology*, Vol. 45, No. 10, pp. 169-174.

Holt, C.J., R.G.S. Matthew, and E. Terzis (1997). A comparative study using the anaerobic baffled reactor to treat a phenolic wastewater. *Eight International Conference on Anaerobic Digestion*, Vol. 2, pp. 40-47.

Huysmans, A., M. Weemaes, P.A. Fonseca, and W. Verstraete (2001). (short communication) Ozonation of activated sludge in the recycle stream. *Journal of Chemical Technology and Biotechnology*, Vol. 76, pp. 321-324.

Jeison, D. and R. Chamy (1999). Comparison of the behaviour of expanded granular sludge bed (EGSB) and upflow anaerobic sludge blanket (UASB) reactors in dilute and concentrated wastewater treatment. *Water Science and Technology*, Vol. 40, No. 8, pp. 91-97.

Jewell, W.J., M.S. Switzenbaum, and J.W. Morris (1981). Municipal wastewater treatment with the anaerobic attached microbial film expanded bed process. *Journal WPCF*, Vol. 53, No. 4, pp. 482-490.

Kalogo, Y. and W. Verstraete (2000). Technical feasibility of the treatment of domestic wastewater by a CEPS-UASB system. *Environmental Technology*, Vol. 21, No.1, pp. 55-65.

Kamiya, T. and Hirotsuji, J. (1998), New Combined System of Biological Process and Intermittent Ozonation for Advanced Wastewater Treatment, *Water Science and Technology*, Vol. 38, No. 8-9, pp. 145-153.

Kobayashi, H.A., M.K. Stenstrom and R.A. Mah (1983). Treatment of Low Strength Domestic Wastewater Using the Anaerobic Filter. *Water Research*, Vol. 17, No. 8, pp. 903-909.

Lacalle, M.L., S. Villaverde, F. Fdz-Polanco, and P.A. Garcia-Encina (2001). Combined anaerobic/aerobic (UASB + UBAF) system for organic matter and nitrogen removal from a high strength of industrial wastewater. *Water Science and Technology*, Vol. 44, No. 4, pp. 255-262.

Langenhoff, A.A.M. and D.C. Stuckey (2000). Treatment of dilute wastewater using an anaerobic baffled reactor: Effect of low temperature. *Water Research*, Vol. 34, No. 15, pp. 3867-3875.

Langenhoff, A.A.M., N. Intrachandra, and D.C. Stuckey (2000). Treatment of dilute soluble and colloidal wastewater using an anaerobic baffled reactor: Influence of hydraulic retention time. *Water Research*, Vol. 34, No. 4, pp. 1307-1317.

Lapara, T.M. and J.E. Alleman (1999). Thermophilic aerobic biological wastewater treatment. *Water Research*, Vol. 33, No. 4, pp. 895-908.

Lapara, T.M., A. Konopka, and J.E. Alleman (2000). Energy spilling by thermophilic aerobes in potassium-limited continuous culture. *Water Research*, Vol. 34, No. 10, pp. 2723-2726.

Lee, J., W.Y. Ahn, and C.H. Lee (2001a). Comparison of the filtration characteristics between attached and suspended growth microorganisms in submerged membrane bioreactor. *Water Research* , Vol. 35, No. 10, pp. 2435-2445.

Lee, N.M. and T. Welander (1996a). Reducing sludge production in aerobic wastewater treatment through manipulation of the ecosystem. *Water Research*, Vol. 30, No. 8, pp. 1781-1790.

Lee, N.M. and T. Welander (1996b). Use of protozoa and metazoa for decreasing sludge production in aerobic wastewater treatment. *Biotechnology Letters*, Vol. 18, No. 4, pp.429-434.

Lee, S.J., K.H. Choo, and C.H. Lee (2000). Conjunctive use of ultrafiltration with powdered activated carbon adsorption for removal of synthetic and natural organic matter. *Journal of Industrial and Engineering Chemistry,* Vol. 6, No. 6, pp. 357-364.

Lee, S.M., J.Y. Jung, and Y.C. Chung (2001b). Novel method for enhancing permeate flux of submerged membrane system in two-phase anaerobic reactor. *Water Research*, Vol. 35, No. 2, pp. 471-477.

Lettinga, G. (2001). Digestion and degradation, air for life. *Water Science and Technology*, Vol. 44, No. 8, pp. 157-176.

Lettinga, G., A.F.M. van Velsen, S.W. Hobma, W.J. de Zeeuw, and A. Klapwijk (1980). Use of the upflow sludge blanket (USB) reactor concept for biological wastewater treatment. *Biotechnology and Bioengineering*, Vol. 22, pp. 699-734.

Liu, W.-T., T. Mino, K. Nakamura, and T. Matsuo (1996) Glycogen accumulating population and its anaerobic substrate uptake abilities in anaerobic-aerobic activated sludge without biological phosphorus removal. *Water Research*, Vol. 30, pp. 75-87.

Liu, W.T, K. Nakamura, T. Matsuo, and T. Mino (1997) Internal energy-based competition between polyphosphate and glycogen-accumulating bacteria in biological phosphorus removal reactors—effect of P/C feeding ratio. *Water Research*, Vol. 31, No. 6, pp.1430-1438.

Liu, Y. (1998). Energy uncoupling in microbial growth under substrate-sufficient conditions. *Applied Microbiology and Biotechnology*, Vol. 49, No. 5, pp. 500-505.

Liu, Y. (2000). Effect of chemical uncoupler on the observed growth yield in batch culture of activated sludge. *Water Research*, Vol. 34, No. 7, pp. 2025-2030.

Liu, Y. (2003). Chemically reduced excess sludge production in the activated sludge process. *Chemosphere*, Vol. 50, No.1, pp. 1-7.

Liu, Y.H., C. Geiger, and A.A. Randall (2002). The role of poly-hydroxy-alkanoate form in determining the response of enhanced biological phosphorus removal biomass to volatile fatty acids. *Water Environment Research*, Vol.74, No. 57, pp. 57-67.

Low, E.W., H.A. Chase, M.G. Milner, and T.P. Curtis (2000). Uncoupling of metabolism to reduce biomass production in the activated sludge process. *Water Research*, Vol. 34, No. 12, pp. 3204-3212.

Macarie, H. (2000). Overview of the application of anaerobic treatment to chemical and petrochemical wastewaters. *Water Science and Technology*, Vol. 42, No. 5-6, pp. 201-213.

Machdar, I., Y. Sekiguchi, H.Sumino, A. Ohashi, and H. Harada (2000). Combination of a UASB reactor and a curtain type DHS (downflow hanging sponge) reactor as a cost-effective sewage treatment system for developing countries. *Water Science and Technology*, Vol. 42, No. 3-4, pp. 83-88.

Madigan, M.T., J.M. Martinko, and J. Parker (1997). *Brock, Biology of Microorganisms,* 8th Edition, Prentice Hall, N.J.

Manariotis, I.D. and S.G. Grigoropoulos (2002). Low-strength wastewater treatment using an anaerobic baffled reactor. *Water Environment Research*, Vol. 74, No. 2, pp. 170-176.

Maragno, A. and J.R. Campos (1992). Treatment of Waste-Water with a Low Concentration of Organics Using an Anaerobic Fluidized-Bed Reactor. *Water Science and Technology*, Vol. 25, No. 7, pp. 179-191.

Matsushige, K.,Y. Inamori, M. Mizuochi, M. Hosomi, and R. Sudo (1990). The Effects of Temperature on Anaerobic Filter Treatment for Low-Strength Organic Waste-Water. *Environmental Technology*, Vol. 11, No. 10, pp. 899-910.

Mayhew, M. and T. Stephenson (1998). Biomass yield reduction: Is biochemical manipulation possible without affecting activated sludge process efficiency? *Water Science and Technology*, Vol. 38, No. 8-9, pp. 137-144.

McCarty, P.L. and D.P. Smith (1986). Anaerobic wastewater treatment. *Environmental Science and Technology*, Vol. 23, pp. 1200-1226.

McClintock, S.A., J.H. Sherrard, J.J. Novak, and C.W. Randall (1988). Nitrate versus oxygen respiration in the activated sludge process. *Journal of Water Pollution Control Federation,* Vol. 60, p. 342.

Mendonca, A.F., T.L. Amoroso, and S.J. Knabel (1994) Destruction of gram-negative food-borne pathogens by high pH involves disruption of the cytoplasmic membrane. *Applied and Environmental Microbiology*, Vol. 60, No. 11, pp. 4009-4014.

Mines, R.O. and R.S. Kalch (1999). Sludge minimization using aerobic/anoxic treatment technology. *In Proceedings of the ASCE-CSCE National Conference on Environmental Engineering,* July 25-28, 1999, Norfolk, VA, pp. 506-515.

Monroy, O., A. Noyola, F. Raminez, and J.P. Guiot (1988). Anaerobic digestion of water hyacinth as a highly efficient treatment process for developing countries. *Fifth International Symposium on Anaerobic Digestion*, Bologna, Italy, pp. 747-757.

Müller, J.A. (2000). Pretreatment processes for the recycling and reuse of sewage sludge. *Water Science and Technology,* Vol. 42, No. 9, pp. 167-174.

Nachaiyasit, S. and D.C. Stuckey (1995). Microbial Response to Environmental Changes in an Anaerobic Baffled Reactor (Abr). *Antonie Van Leeuwenhoek International Journal of General and Molecular Microbiology*, Vol. 67, No. 1, pp. 111-123.

Nachaiyasit, S. and D. C.Stuckey (1997a). Effect of low temperatures on the performance of an anaerobic baffled reactor (ABR). *Journal of Chemical Technology and Biotechnology*, Vol. 69, No. 2, pp. 276-284.

Nachaiyasit, S. and D.C. Stuckey (1997b). The effect of shock loads on the performance of an anaerobic baffled reactor (ABR).1. Step changes in feed concentration at constant retention time. *Water Research*, Vol. 31, No. 11, pp. 2737-2746.

Nachaiyasit, S. and D.C. Stuckey (1997c). The effect of shock loads on the performance of an anaerobic baffled reactor (ABR).2. Step and transient hydraulic shocks at constant feed strength. *Water Research*, Vol. 31, No. 11, pp. 2747-2754.

Ndon, U.J. and R.R. Dague (1997a). Ambient temperature treatment of low strength wastewater using anaerobic sequencing batch reactor. *Biotechnology Letters*, Vol. 19, No. 4, pp. 319-323.

Ndon, U.J. and R.R. Dague (1997b). Effects of temperature and hydraulic retention time on anaerobic sequencing batch reactor treatment of low-strength wastewater. *Water Research*, Vol. 31, No. 10, pp. 2455-2466.

Nolasco, M.A., A.L.O. Campos, A.M. Springer, and E.C. Pires (2002). Use of lysis and recycle to control excess sludge production in activated sludge treatment: bench scale study and effect of chlorniated organic compounds. *Water Science and Technology*, Vol. 46, No. 10, pp. 55-61.

Okey, R.W. and H.D. Stensel (1993). Uncouplers and activated sludge–the impact on synthesis and respiration. *Toxicological and Environmental Chemistry*, Vol. 40, No. 1-4, pp. 235-254.

Orozco, A. (1997). Pilot and full-scale anaerobic treatment of low strength wastewater at suboptimal temperature (15 °C) with a hybrid plug flow reactor. *Eighth International Conference on Anaerobic Digestion*, Sendai, Japan, 2, pp. 183-191.

Parkin, G.F., M.A. Sneve, and H. Loos (1991). Anaerobic Filter Treatment of Sulfate-Containing Wastewaters. *Water Science and Technology*, Vol. 23, No. 7-9, pp. 1283-1291.

Parmar, N., A. Singh, and O.P. Ward (2001). Enzyme treatment to reduce solids and improve settling of sewage sludge. *Journal of Industrial Microbiology and Biotechnology*, Vol. 26, pp. 383-386.

Penaud, V., J.P. Delgenes, and R. Moletta (1999). Thermo-chemical pretreatment of a microbial biomass: influence of sodium hydroxide addition on solubilization and anaerobic biodegradability. *Enzyme and Microbiol Technology*, Vol. 25, pp. 258-263.

Polprasert, C., P. Kemmadamrong, and F.T. Tran (1992). Anaerobic Baffle Reactor (Abr) Process for Treating a Slaughterhouse Waste-Water. *Environmental Technology*, Vol. 13, No. 9, pp. 857-865.

Raman, V. and A.N. Khan (1978). Outflow anaerobic filter: a simple sewage treatment device. *International Conference and Water Pollution Control in Developing Countries*, pp. 639-648.

Randall, C.W., K.P. Brennen, S.A. McClintock, and V.M. Pattarkine (1992). The case of anaerobic reduction of oxygen requirements in biological phosphorus removal systems. *Water Environment Research*, Vol. 64, No. 6, pp. 824-833.

Rebac, S. (1998). Psychrophilic anaerobic treatment of low strength wastewaters. Wageningen, Netherlands, Wageningen Agricultural University.

Rensink, J.H. and W.H. Rulkens (1997). Using metazoa to reduce sludge production. *Water Science and Technology*, Vol. 36, No. 11, pp. 171-179.

Rich, L.G. and O.W. Yates (1955). The effect of 2,4-dinitrophenol on activated sludge. *Journal of Applied Microbiology,* Vol. 3, pp. 95-98.

Rocher, M., G. Goma, A. Pilas Begue, L. Louvel, and J.L. Rols (1999). Towards a reduction in excess sludge production in activated sludge processes: biomass physicochemical treatment and biodegredation. *Applied Microbiology and Biotechnology*, Vol. 51, pp. 883-890.

Roels, J. and W. Verstraete (2001). Biological formation of volatile phosphorus compounds. *Bioresource Technology*, Vol. 79, No. 3, pp. 243-250.

Saby, S., M. Djafer, and G.-H. Chen (2002). Feasibility of using a chlorination step to reduce excess sludge in activated sludge process. *Water Research*, Vol. 36, pp. 656-666.

Sakai, Y., T. Fukase, H. Vasui, and H. Shibata (1997). An Activated Sludge Process Without Excess Sludge Production. *Water Science and Technology*, Vol. 36, No. 11, pp. 163-170.

Sakai, Y., T. Aoyagi, N. Shiota, A. Akasahi, and S. Hasegawa (2000). Complete decomposition of biological waste sludge by thermophilic aerobic bacteria. *Water Science and Technology*, Vol. 42, No. 9, pp. 81-88.

Sanz, I. and F. Fdzpolanco (1990). Low-Temperature Treatment of Municipal Sewage in Anaerobic Fluidized-Bed Reactors. *Water Research*, Vol. 24, No. 4, pp. 463-469.

Sayed, S.K.I. and M.A.A. Fergala (1995). Two-stage UASB concept for treatment of domestic sewage including sludge stabilization process. *Water Science and Technology*, Vol. 32, No. 11, pp. 55-63.

Soedjono, E.S. and C.F. Forster (1999). The treatment of simulated sewage by alternating anaerobic double filtration (AADF). *Environmental Technology*, Vol. 20, No. 3, pp.337-342.

Springer, A.M. and J. Higgins, (1999). Mechanical cell lysis and recycle to reduce waste activated sludge. *Proceedings of the TAPPI International Environmental Conference*, Nashville, TN, U.S.A.

Springer, A.M., G. Dietrich-Velazquez, C.M. Higby, and D. Digiacomo (1996). Feasibility study of sludge lysis and recycle in the activated sludge process. *TAPPI*, Vol. 79, No. 5, pp. 162-170.

Stephens, H.L. and H.D. Stensel (1998). Effect of Operating Conditions on Biological Phosphorus Removal. *Water Environment Research*, Vol. 70, No. 3, pp. 362-369.

Stephenson, R.J. and H.S. Dhaliwal (1998). Method of liquefying microorganisms derived from biological wastewater treatment processes. *U.S. Patent No. 6,013,183*.

Strand, S.E., G.N. Harem, and H.D. Stensel (1999). Activated Sludge Yield Reduction Using Chemical Uncouplers. *Water Environment Research*, Vol. 71, pp. 454-458.

Stuart, P., R. Kenny, and J. Sointio (2001). A critical review of the low sludge production (LSP) process for the pulp and paper industry. *Pulp & Paper-Canada*, Vol. 102, No. 4, pp. 23-27.

Switzenbaum, M.S. and W.J. Jewell (1978). Anaerboic attached film expanded bed reactor treatment of dilute organics. *Proceedings of the 51st Annual WPCF Conference*, Anaheim CA, WPCF, pp. 1-164.

Tay, J.H. and X.Y. Zhang (2000). Stability of high-rate anaerobic systems. I: Performance under shocks. *Journal of Environmental Engineering-ASCE,* Vol. 126, No. 8, pp.713-725.

Tchobanoglous, G., F.L. Burton, and H.D. Stensel (2003). *Wastewater Engineering Treatment and Reuse, Fourth Edition*. Metcalf and Eddy, McGraw Hill. New York City, N.Y., USA.

Tiehm, A., K. Nickel, M. Zellhorn, and U. Neis (2001). Ultrasonic waste activated sludge disintegration for improving anaerobic stabilization. *Water Research*, Vol. 35, No. 8, pp. 2003-2009.

Tilche, A., G. Bortone, G. Garuti, and F. Malaspina (1996). Post-treatments of anaerobic effluents. *Antonie Van Leeuwenhoek International Journal of General and Molecular Microbiology*, Vol. 69, No. 1, pp. 47-59.

Torres, P. and E. Foresti (2001). Domestic sewage treatment in a pilot system composed of UASB and SBR reactors. *Water Science and Technology*, Vol. 44, No. 4, pp. 247-253.

Ueda, T. and K. Hata (1999). Domestic wastewater treatment by a submerged membrane bioreactor with gravitational filtration. *Water Research,* Vol. 33, No. 12, pp. 2888-2892.

Uemura, S. and H. Harada (2000). Treatment of sewage by a UASB reactor under moderate to low temperature conditions. *Bioresource Technology*, Vol. 72, No. 3, pp. 275-282.

Vanderlast, A.R.M. and G. Lettinga (1992). Anaerobic Treatment of Domestic Sewage under Moderate Climatic (Dutch) Conditions Using Upflow Reactors at Increased Superficial Velocities. *Water Science and Technology*, Vol. 25, No. 7, pp.167-178.

van Lier, J.B., F.P. van der Zee, N.C.G. Tan, S. Rebac, and R. Kleerebezem (2001). Advances in high-rate anaerobic treatment: staging of reactor systems. *Water Science and Technology,* Vol. 44, No. 8, pp. 15-25.

Van Munch, E. (1998). DSP Prefermenter Technology Book, in Liu et al. (2003)

von Sperling, M., V.H. Freire, and C.A.D. Chernicharo (2001). Performance evaluation of a UASB–activated sludge system treating municipal wastewater. *Water Science and Technology*, Vol. 43, No. 11, pp. 323-328.

Wable, M.V. and C.W. Randall (1994). Investigation of hypothesized anaerobic stabilization mechanisms in biological nutrient removal systems. *Water Environment Research*, Vol. 66, No. 2, pp. 101-167.

Wadehra, P., C. Henson, Sr., J. Packman, E. Singleton, J. Berger, J. Benson, J. Urbanik, and W.D. Jellis (1999). Reduced wasting from activated sludge processes using a mechanical cell lysis technology. *Proceedings of the Water Environment Federation Annual Conference*, New Orleans, LA, U.S.A.

Wang, K. (1994). Integrated anaerobic and anaerobic treatment of sewage. *Sub-Dept. Environmental Technology*, Wageningen University, NL.

Water Environment Federation (2000). *Aerobic Fixed-Growth Reactors: A Special Publication*, Water Environment Federation, Alexandria, VA.

Welander, T., T. Alexandersson, T. Ericsson, L. Gunnarsson, and A. Storlie (2000). Reducing sludge production in biological effluent treatment by applying the LSP process. *TAPPI International Environmental Conference Exhibition*, Denver, 6-10 May, 2000, Book 2, pp.757-764.

Welander, T., L.E. Olsson, and C. Fasth (2002). Nutrient-limited biofilm pretreatment: an efficient way to upgrade activated sludge plants. *Tappi Journal*, Vol. 1, No. 4, pp. 20-26.

Wen, C., X. Huang, and Y. Qian (1999). Domestic wastewater treatment using an anaerobic bioreactor coupled with membrane filtration. *Process Biochemistry*, Vol. 35, No. 3-4, pp. 335-340.

Witthauer, D. and D.C. Stuckey (1982). Laboratory studies on anaerobic processes to treat dilute organic waste water in developing countries. Dubendorf, Switzerland, IRCWD, EAWAG.

Xing, J. and A. Tilche (1992). The Effect of Hydraulic Retention Time on the Hybrid Anaerobic Baffled Reactor Performance at Constant Loading. *Biomass & Bioenergy*, Vol. 3, No. 1, pp. 25-29.

Xing, J., R. Boopathy, and A. Tilche (1991). Model Evaluation of Hybrid Anaerobic Baffled Reactor Treating Molasses Waste-Water. *Biomass & Bioenergy*, Vol. 1, No. 5, pp. 267-274.

Yang, P.Y. and T.H. Moengangongo (1987). Operational Stability of a Horizontally Baffled-Anaerobic Reactor for Diluted Swine Waste-Water in the Tropics. *Transactions of the ASAE*, Vol. 30, No. 4, pp. 1105-1110.

Yang, X., G. Garuti, R. Farina, V. Parisi, and A. Tilche (1988). Process differences between a sludge bed filter and an anaerobic baffled reactor treating soluble wastes. *Fifth International Symposium on Anaerobic Digestion,* Bologna, Italy, pp. 335-360.

Yasui, H. and Shibata, M. (1994). An innovative approach to reduce excess sludge production in the activated sludge process. *Water Science and Technology*, Vol. 30, No. 9, pp. 11-20.

Yasui, H., K. Nakamura, S. Sakuma, M. Iwasaki, and Y. Sakai (1996). A Full-Scale Operation of a Novel Activated Sludge Process Without Excess Sludge Production, *Water Science and Technology*, Vol. 14, No. 3-4, pp. 395-404.

Zehnder, A.J.B. (1988). *Biology of Anaerobic Microorganisms*, John Wiley and Sons, New York City, NY, U.S.A.

Zellner, G., F. Neudorfer, and H. Diedmann (1994). Degradation of lactate by an anaerobic mixed culture in a fluidized-bed reactor. *Water Research*, Vol. 28, No. 6, pp. 1337-1340.

WASTEWATER UTILITY

Alabama
Montgomery Water Works & Sanitary Sewer Board

Alaska
Anchorage Water & Wastewater Utility

Arizona
Gila Resources

Glendale, City of, Utilities Department

Mesa, City of

Peoria, City of

Phoenix Water Services Department

Pima County Wastewater Management

Arkansas
Little Rock Wastewater Utility

California
Calaveras County Water District

Central Contra Costa Sanitary District

Contra Costa Water District

Crestline Sanitation District

Delta Diablo Sanitation District

Dublin San Ramon Services District

East Bay Dischargers Authority

East Bay Municipal Utility District

El Dorado Irrigation District

Fairfield-Suisun Sewer District

Fresno Department of Public Utilities

Irvine Ranch Water District

Las Virgenes Municipal Water District

Livermore, City of

Lodi, City of

Los Angeles, City of

Los Angeles County, Sanitation Districts of

Napa Sanitation District

Orange County Sanitation District

Palo Alto, City of

Riverside, City of

Sacramento Regional County Sanitation District

San Diego Metropolitan Wastewater Department, City of

San Francisco, City & County of

San Jose, City of

Santa Barbara, City of

Santa Rosa, City of

South Bayside System Authority

South Coast Water District

South Orange County Wastewater Authority

Union Sanitary District

West Valley Sanitation District

Colorado
Boulder, City of

Colorado Springs Utilities

Littleton/Englewood Water Pollution Control Plant

Metro Wastewater Reclamation District, Denver

Connecticut
The Mattabasset District

New Haven, City of, WPCA

District of Columbia
District of Columbia Water & Sewer Authority

Florida
Broward, County of

Fort Lauderdale, City of

Gainesville Regional Utilities

JEA

Miami-Dade Water & Sewer Authority

Orange County Utilities Department

Orlando, City of

Reedy Creek Improvement District

Seminole County Environmental Services

St. Petersburg, City of

Stuart Public Utilities

Tallahassee, City of

Tampa, City of

Toho Water Authority

West Palm Beach, City of

Georgia
Atlanta Department of Watershed Management

Augusta, City of

Clayton County Water Authority

Cobb County Water System

Columbus Water Works

Fulton County

Gwinnett County Department of Public Utilities

Macon Water Authority

Savannah, City of

Hawaii
Honolulu, City and County of

Illinois
American Bottoms Wastewater Treatment Plant

Greater Peoria Sanitary District

Kankakee River Metropolitan Agency

Metropolitan Water Reclamation District of Greater Chicago

North Shore Sanitary District

Wheaton Sanitary District

Indiana
Fort Wayne, City of

Iowa
Ames, City of

Cedar Rapids Wastewater Facility

Des Moines Metro Wastewater Reclamation Authority

Iowa City, City of

Kansas
Johnson County Unified Wastewater Districts

Unified Government of Wyandotte County/Kansas City, City of

Kentucky
Louisville & Jefferson County Metropolitan Sewer District

Louisiana
Sewerage & Water Board of New Orleans

Maine
Bangor, City of

Portland Water District

Maryland
Anne Arundel County Bureau of Utility Operations

Howard County Department of Public Works

Washington Suburban Sanitary Commission

Massachusetts
Boston Water & Sewer Commission

Upper Blackstone Water Pollution Abatement District

Michigan
Ann Arbor, City of

Detroit, City of

Holland Board of Public Works

Lansing, City of

Owosso Mid-Shiawassee County WWTP

Saginaw, City of

Wayne County Department of Environment

Wyoming, City of

Minnesota
Rochester, City of

Western Lake Superior Sanitary District

Missouri
Independence, City of

Kansas City Missouri Water Services Department

Little Blue Valley Sewer District

Nebraska
Lincoln Wastewater System

Nevada
Henderson, City of

New Jersey
Passaic Valley Sewerage Commissioners

New York
New York City Department of Environmental Protection

Rockland County Solid Waste Management Authority/ Sewer District

North Carolina
Cary, Town of

Charlotte/Mecklenburg Utilities

Durham, City of

Metropolitan Sewerage District of Buncombe County

Orange Water & Sewer Authority

Ohio
Akron, City of

Butler County Department of Environmental Services

Columbus, City of

Metropolitan Sewer District of Greater Cincinnati

Northeast Ohio Regional Sewer District

Summit, County of

Oklahoma
Tulsa, City of

Oregon
Clean Water Services

Eugene/Springfield Water Pollution Control

Water Environment Services

Pennsylvania
Philadelphia, City of

University Area Joint Authority, State College

South Carolina
Charleston Commissioners of Public Works

Mount Pleasant Waterworks & Sewer Commission

Spartanburg Sanitary Sewer District

Tennessee
Cleveland, City of

Knoxville Utilities Board

Murfreesboro Water & Sewer Department

Nashville Metro Water Services

Texas
Austin, City of

Dallas Water Utilities

Denton, City of

El Paso Water Utilities

Fort Worth, City of

Gulf Coast Waste Disposal Authority

Houston, City of

San Antonio Water System

Trinity River Authority

Utah
Salt Lake City Corporation

Virginia
Alexandria Sanitation Authority

Arlington, County of

Fairfax County Virginia

Hampton Roads Sanitation District

Henrico, County of

Hopewell Regional Wastewater Treatment Facility

Loudoun County Sanitation Authority

Lynchburg Regional WWTP

Prince William County Service Authority

Richmond, City of

Rivanna Water & Sewer Authority

Washington
Edmonds, City of

Everett, City of

King County Department of Natural Resources

Seattle Public Utilities

Sunnyside, Port of

Yakima, City of

Wisconsin
Green Bay Metro Sewerage District

Madison Metropolitan Sewerage District

Milwaukee Metropolitan Sewerage District

Racine, City of

Sheboygan Regional Wastewater Treatment

Wausau Water Works

Australia
South Australian Water Corp.

Sydney Water Corp.

Water Corp. of Western Australia

Canada
Greater Vancouver Regional District

Toronto, City of, Ontario

Winnipeg, City of, Manitoba

Mexico
Servicios de Agua y Drenaje de Monterrey, I.P.D.

New Zealand
Watercare Services Limited

Singapore
Singapore Public Utilities Board

United Kingdom
Yorkshire Water Services Limited

STORMWATER UTILITY

California
Los Angeles, City of, Department of Public Works

Monterey, City of

Santa Rosa, City of

Colorado
Boulder, City of

Georgia
Griffin, City of

Iowa
Cedar Rapids Wastewater Facility

Des Moines Metro Wastewater Reclamation Authority

Kansas
Overland Park, City of

Kentucky
Louisville & Jefferson County Metropolitan Sewer District

Maine
Portland Water District

Minnesota
Western Lake Superior Sanitary District

North Carolina
Cary, Town of

Charlotte, City of, Stormwater Services

Pennsylvania
Philadelphia, City of

Tennessee
Chattanooga Storm Water Management

Washington
Seattle Public Utilities

STATE

Arkansas Department of Environmental Quality

Fresno Metropolitan Flood Control District, Calif.

Urban Drainage & Flood Control District, Colo.

CORPORATE

ADS Environmental Services

The ADVENT Group Inc.

Alan Plummer & Associates

Alden Research Laboratory

Alpine Technology Inc.

Aqua-Aerobic Systems, Inc.

Aquateam–Norwegian Water Technology Centre A/S

BaySaver Inc.

BioVir Laboratories, Inc.

Black & Veatch

Boyle Engineering Corporation

BPR CSO

Brown & Caldwell

Burns & McDonnell

CABE Associates Inc.

The Cadmus Group

Camp Dresser & McKee Inc.

Carollo Engineers Inc.

Carpenter Environmental Associates Inc.

CDS Technologies Inc.

Chemtrac Systems Inc.

CH2M HILL

Damon S. Williams Associates, LLC

David L. Sheridan, P.C.

Earth Tech Inc.

Eco-Matrix

Environmental Engineers International

EMA Inc.

The Eshelman Company Inc.

Finkbeiner, Pettis, & Strout Inc.

Freese & Nichols, Inc.

Frontier Geosciences Inc.

ftn Associates Inc.

Fuss & O'Neill, Inc.

Gannett Fleming Inc.

Golder Associates Inc.

Greeley and Hansen LLC

Hazen & Sawyer, P.C.

HDR Engineering Inc.

HNTB Corporation

HydroQual Inc.

Ingersoll-Rand Energy Systems

Insituform Technologies Inc.

Institute for Environmental Technology & Industry, Korea

Jacobson Helgoth Consultants Inc.

Jason Consultants Inc.

Jordan, Jones, & Goulding Inc.

KCI Technologies Inc.

Kelly & Weaver, P.C.

Kennedy/Jenks Consultants

Komline Sanderson Engineering Corporation

Lawler, Matusky & Skelly Engineers, LLP

Limno-Tech Inc.

Lombardo Associates Inc.

Malcolm Pirnie Inc.

McKim & Creed

MEC Analytical Systems Inc.

Metcalf & Eddy Inc.

MWH

New England Organics

Odor & Corrosion Technology Consultants Inc. (OCTC)

ONDEO Degremont Inc.

Oswald Green, LLC

PA Government Services Inc.

Parametrix Inc.

Parsons

Post, Buckley, Schuh & Jernigan

R&D Engineering/Conestoga Rover & Associates

The RETEC Group

RMC, Inc. (Raines, Melton & Carella)

R.M. Towill Corporation

Ross & Associates Ltd.

Royce Technologies

SAIC Maritime Technical Group

Sear-Brown

Short Elliott Hendrickson, Inc.

Stormwater Management, Inc.

Synagro Technologies, Inc.

Tetra Tech Inc.

Trojan Technologies Inc.

URS Corporation

Veolia Water NATC

Wade-Trim Inc.

Weston Solutions Inc.

Woodard & Curran

WRc/D&B, LLC

WWETCO, LLC

Zoeller Pump Company

INDUSTRY

American Electric Power

ChevronTexaco Energy Research & Technology Company

The Coca-Cola Company

Dow Chemical Company

DuPont Company

Eastman Kodak Company

Merck & Company Inc.

ONDEO Services

Procter & Gamble Company

PSEG Services Corp.

Severn Trent Services Inc.

Shell Global Solutions

Thames Water Plc

United Water Services LLC

Note: List as of 1/31/04

WERF Product Order Form

As a benefit of joining the Water Environment Research Foundation, subscribers are entitled to receive one complimentary copy of all final reports and other products.Additional copies are available at cost (usually $10).To order your complimentary copy of a report, please write "free" in the unit price column.WERF keeps track of all orders.If the charge differs from what is shown here, we will call to confirm the total before processing.

Name Title

Organization

Address

City State Zip Code Country

Phone Fax Email

Stock #	Product	Quantity	Unit Price	Total

Postage & Handling

VA Residents Add 4.5% Sales Tax

Canadian Residents Add 7% GST

TOTAL

Method of Payment:

(All orders must be prepaid.)

❑ Check or Money Order Enclosed

❑ Visa ❑ Mastercard ❑ American Express

Account No. Exp. Date

Signature

Shipping & Handling:

Amount of Order	United States	Canada & Mexico	All Others
Up to but not more than:	Add:	Add:	Add:
$20.00	$5.00*	$8.00	50% of amount
30.00	5.50	8.00	40% of amount
40.00	6.00	8.00	
50.00	6.50	14.00	
60.00	7.00	14.00	
80.00	8.00	14.00	
100.00	10.00	21.00	
150.00	12.50	28.00	
200.00	15.00	35.00	
More than $200.00	Add 20% of order	Add 20% of order	

*minimum amount for all orders

Note: Please make checks payable to the Water Environment Research Foundation.

To Order (Subscribers Only):

Log on to www.werf.org and click on the "Product Catalog."

Phone: (703) 684-2470
Fax: (703) 299-0742.

WERF
Attn: Subscriber Services
635 Slaters Lane
Alexandria, VA 22314-1177

To Order (Non-Subscribers):

Non-subscribers may be able to order WERF publications either through WEF (www.wef.org) or IWAP (www.iwapublishing.com).Visit WERF's website at www.werf.org for details.

IWA Publishing's authorised EU representative for General Product Safety Regulations is Diane D'Arras, 15 rue Duret, 75116 Paris, France, e-mail: safety@iwap.co.uk.

Printed and bound by CPI Group (UK) Ltd, Croydon, CR0 4YY
12/05/2026
02108865-0004